THE ILLUSTRATED GUIDE TO THE SOLAR SYSTEM

ALEXANDER GORDON SMITH

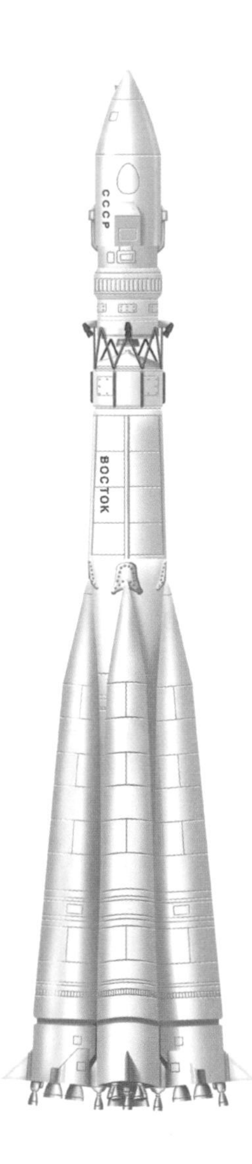

This edition is published by Southwater,
an imprint of Anness Publishing Ltd,
Hermes House, 88–89 Blackfriars Road,
London SE1 8HA; tel. 020 7401 2077;
fax 020 7633 9499

www.southwaterbooks.com;
www.annesspublishing.com

Anness Publishing has a new picture agency
outlet for images for publishing, promotions
or advertising. Please visit our website
www.practicalpictures.com for more information.

UK agent: The Manning Partnership Ltd;
tel. 01225 478444; fax 01225 478440;
sales@manning-partnership.co.uk
UK distributor: Book Trade Services;
tel. 0116 2759086; fax 0116 2759090;
uksales@booktradeservices.com;
exportsales@booktradeservices.com
North American agent/distributor:
National Book Network; tel. 301 459 3366;
fax 301 429 5746; www.nbnbooks.com
Australian agent/distributor:
Pan Macmillan Australia; tel. 1300 135 113;
fax 1300 135 103;
customer.service@macmillan.com.au
New Zealand agent/distributor: David Bateman Ltd;
tel. (09) 415 7664; fax (09) 415 8892

Picture Sources

Atlas Collection: 66–67c, 71tl; **David Blundell:** 13ct, 13t, 14–15b, 20–21c (Image courtesy of NASA/NSSDC), 103tr; **Peter Bull Art Studio:** 16b, 20b, 25t, 35br, 40bl, 44cl, 48cl, 51bl, 53tr, 54bl, 55tr, 68bl, 69br, 73br, 78c, 78–79cb, 98bl, 113tl, 113r; **Robin Carter:** 22–23c, 42–43c; **Stefan Chabluk:** 10bl, 10–11c, 24–25c, 25c, 34bl, 34b (inset), 38–39c, 40–41c, 48–49c, 50–51c; **Corel Picture Library:** 10t, 11br, 21br, 24b, 71br, 76cl, 111br; **Digital Vision:** 14–15c, 18cr, 19cl, 18t, 19b, 29br, 31br, 31cr, 32bl, 32–33c, 33bl, 33br, 26bl, 26–27c, 51cr, 53br, 54-55c, 54tr, 56c, 57tr, 57br, 58tr, 59tr, 59b, 60–61c, 61r (all), 62bl, 62c, 63l, 64–65ct, 78tr, 82cl, 82tr, 83tr (all), 86bl, 88-89c, 90bl, 90–91c, 94bl, 94tr, 95c, 96cl, 97t, 98tr, 100bl, 100cr, 101c, 102tr, 110–111c, 117r, 118l, 118tr; **Mary Evans Picture Library:** 77t, 77br; **NASA/NSSDC:** 18bl, 30bl, 36bl, 43br, 102bl; **NSSDC/Dr R Albrecht/ESA/ESO:** 38cl; **NASA/JPL/Caltech:** 17br, 27b, 30–31c, 34t, 35bl, 35tr, 37ct, 37cb, 49tr, 91tr, 91br; **NASA/R Williams (ST Scl):** 63tr; **Natural History Museum:** 45cr; **Terry Pastor:** 8–9c, 23br, 28–29c, 38cb, 43cb, 36tr, 52bl, 52–53c, 55cr, 64bl, 64br, 65bl, 65cr, 72–73c, 80–81c, 83b, 84–85c, 86c, 86r, 87l, 87tr, 89cr, 92–93c (all), 99br, 107t, 107b, 112–113bc, 116b, 120c, 121c; **Science Photo Library:** 76ct, 77bl, 80bl, Julian Baum/Science Photo Library: 57bl, 104–105t, Chris Butler/Science Photo Library: 41br, Tony Craddock/ Science Photo Library: 68tr, Fred Espenak/Science Photo Library: 16tr, John Foster/Science Photo Library: 28bl, Mark Garlick/Science Photo Library: 104–105c, 107cr, David Hardy/Science Photo Library: 39br, 106–107c, Roger Harris/Science Photo Library: 108c,109b, Dr Michael J Ledlow/Science Photo Library: 17tr, Gregory MacNicol/Science Photo Library: 70bl, NASA/Science Photo Library: 27tr, 93br, 93bc (inset), 95tr, 97br, 99tc, 105cr, NOAO/Science Photo Library: 57cr, Novosti/Science Photo Library: 84bl, 85cr, David Nunuk/Science Photo Library: 12bc, 44tr, 81br, David Parker/Science Photo Library: 44–45c, 87br, 110bl, Rev. Ronald Royer/Science Photo Library: 12–13c, 12bl, Dr Seth Shostack/Science Photo Library: 104bl, US Geological Survey/Science Photo Library: 103br, Victor Habbick Visions/Science Photo Library: 70–71c 109tl, Jason Ware/Science Photo Library: 50tr, Frank Zullo/ Science Photo Library: 13br, 58–59c; **Search for Extraterrestrial Intelligence/Seth Shostak:** 68–69c.

Produced by Nicola Baxter
Designer: Amanda Hawkes

Ethical Trading Policy
Because of our ongoing ecological investment programme, you, as our customer, can have the pleasure and reassurance of knowing that a tree is being cultivated on your behalf to naturally replace the materials used to make the book you are holding. For further information about this scheme, go to www.annesspublishing.com/trees

Previously published as *The Solar System*

Publisher's Note
Although the advice and information in this book are believed to be accurate and true at the time of going to press, neither the authors nor the publisher can accept any legal responsibility or liability for any errors or omissions that may be made.

Contents

Beyond the Solar System · 46

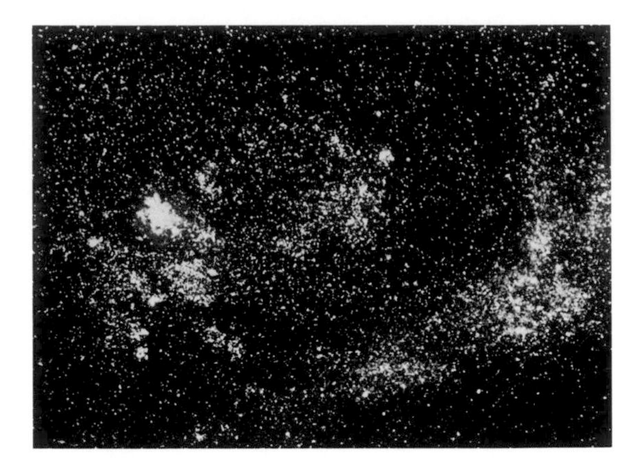

Exploring the Solar System · 74

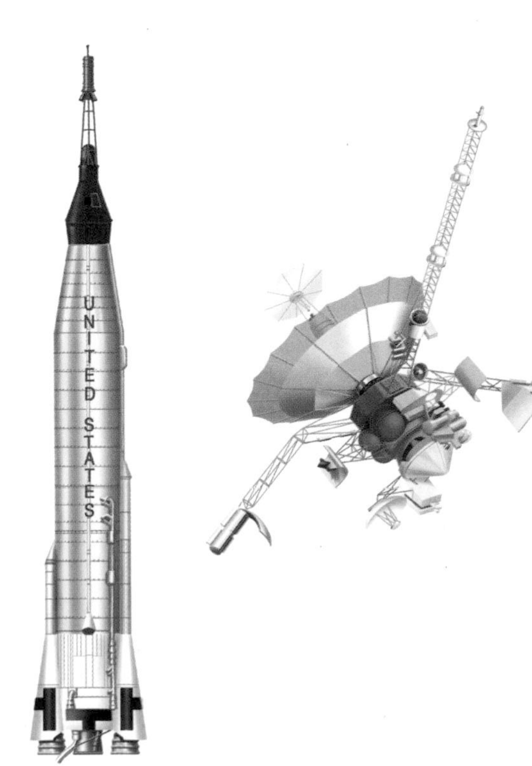

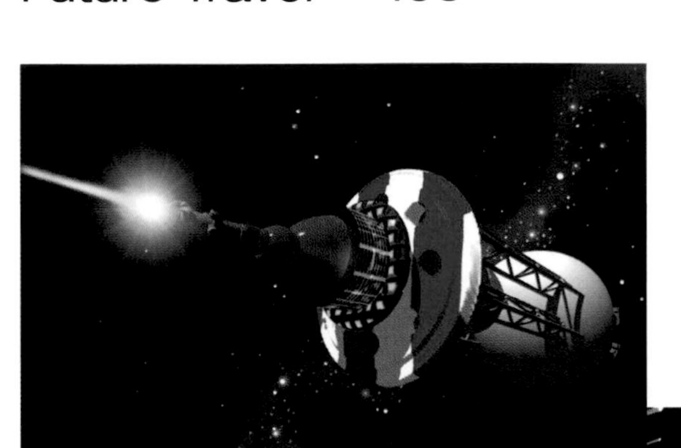

Reference Section · 114

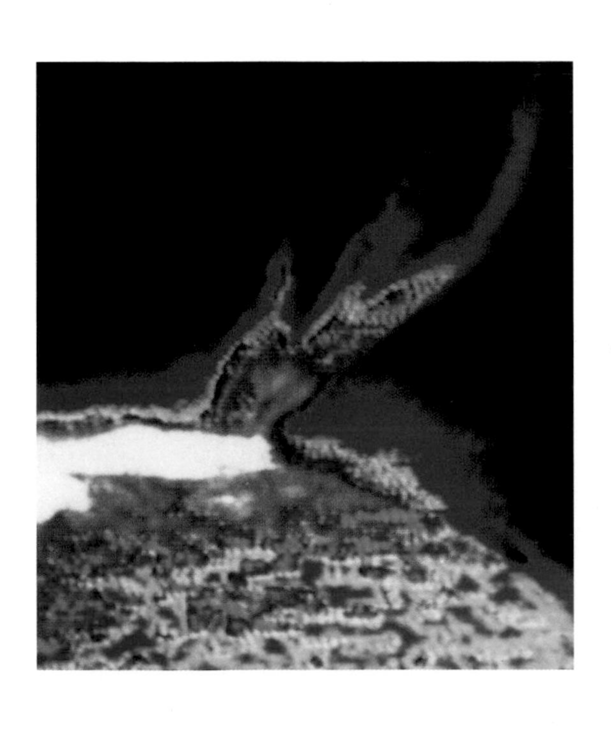

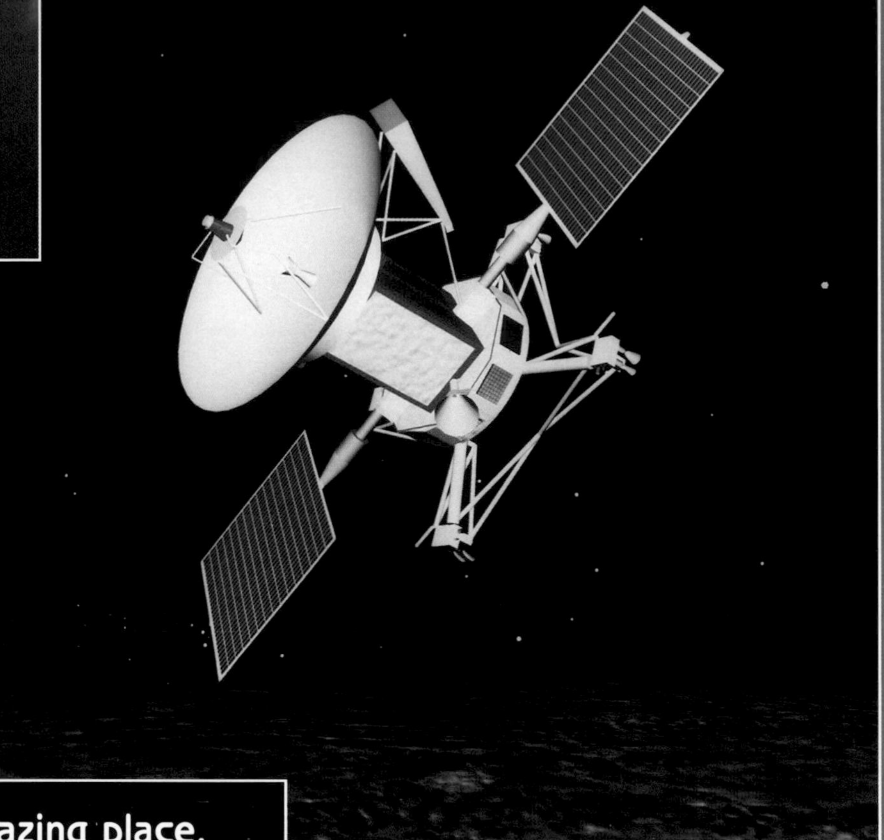

Our Solar System is an amazing place. Our planet, Earth, is one of nine that orbit a giant nuclear furnace called the Sun—a furnace so hot that its core is over 15,000,000°C (32,400,000°F). Apart from Earth, the planets in the Solar System are not welcoming to life. They range from cloud-covered death-traps to enormous balls of gas on which giant storms rage continuously. One has a planet-wide volcano that covers everything with molten lava. Another has winds of over 2,000 kilometres (1,240 miles) per hour. Between the planets, collections of rock speed through space, destroying smaller objects in their paths, and giant moons, some with volcanoes of ice, circle their parent planets at incredible speeds.

Discovering the Solar System

Our Solar System

Formed from the force of an exploding star, our Solar System is a neighbourhood in space. Held in place by the enormous gravitational pull of the Sun are nine planets, over sixty moons and billions of asteroids and comets. These stretch over a distance of 7.6 million million (trillion) kilometres (4.7 trillion miles). Scientists believe there could be countless similar systems in the Universe, but with the nearest possibility being over 43 trillion kilometres (27 trillion miles) away, it is impossible to tell for sure.

Facts and Figures

- Diameter of Solar System: 1.6 light years
- Age: 4.6 billion years
- Time taken to orbit centre of Milky Way: 220 million years
- Star: the Sun
- Total number of planets: 9
- Total number of moons: 63
- Future life: 5 billion years

2. The cloud began to shrink and rotated faster and faster. This movement caused many of the particles in the cloud to group at the centre, where they heated up to form the young star that was to become our Sun.

1. The force of a nearby supernova caused a cloud of gas and dust to rotate and collapse. The supernova enriched the cloud with many different materials, including carbon, which is fundamental to life.

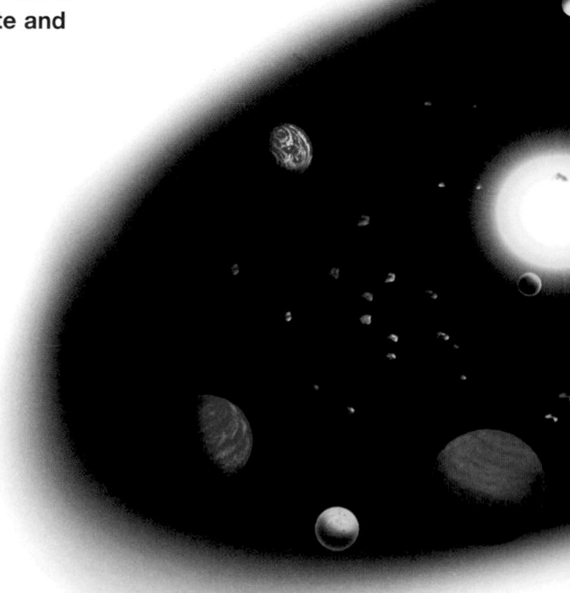

Birth of a solar system

Scientists believe that the Solar System came about because of the death of a star. When some giant stars reach the end of their lives, they explode violently as supernovae, sending shockwaves of energy into space. Around 4.6 billion years ago, one of these shockwaves (picture 1), travelling at over 30 million kilometres (19 million miles) an hour, hit a cloud of gas, dust and ice. The strength of this impact forced the gases to flatten into a swirling spiral of debris (picture 2).

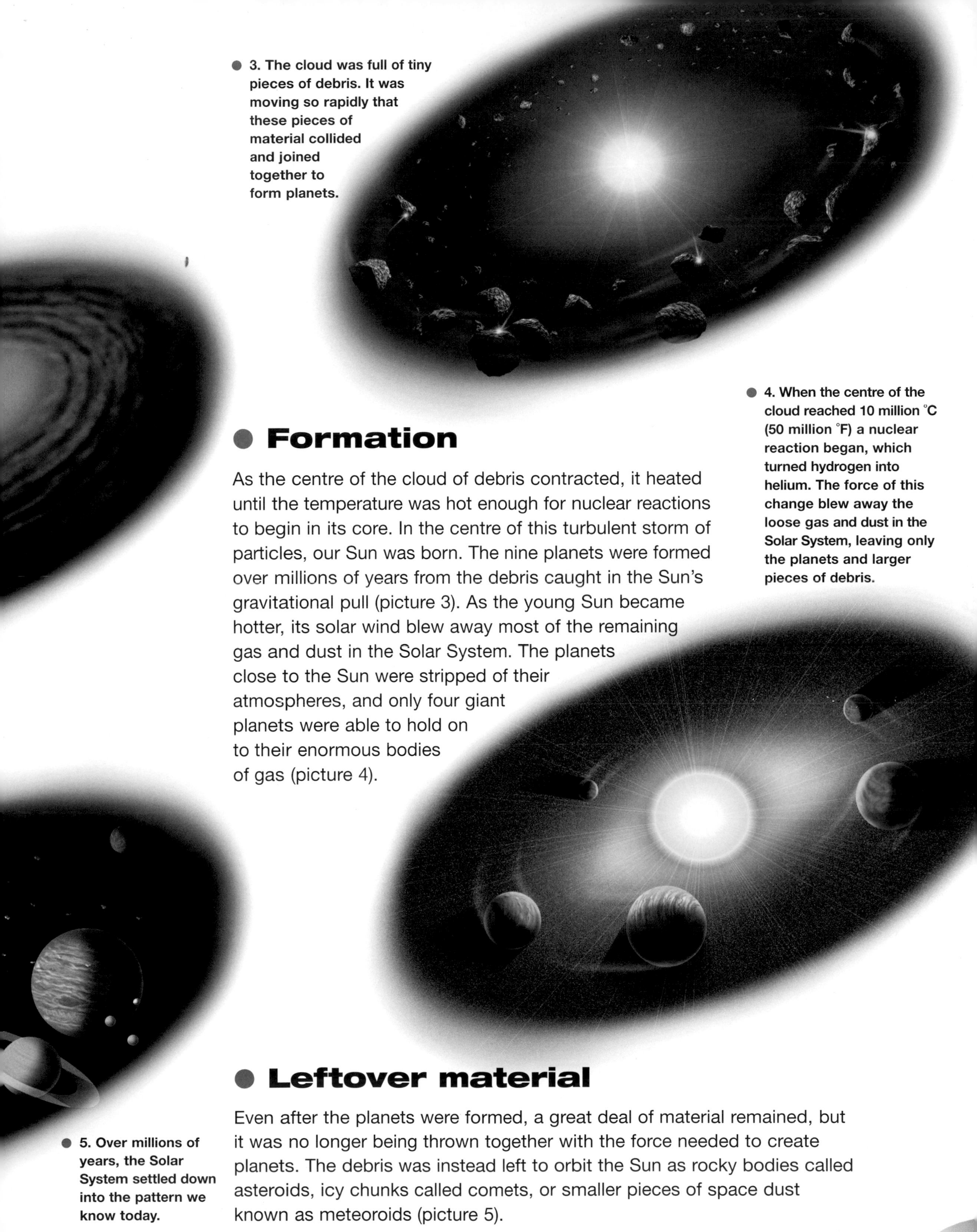

3. The cloud was full of tiny pieces of debris. It was moving so rapidly that these pieces of material collided and joined together to form planets.

Formation

As the centre of the cloud of debris contracted, it heated until the temperature was hot enough for nuclear reactions to begin in its core. In the centre of this turbulent storm of particles, our Sun was born. The nine planets were formed over millions of years from the debris caught in the Sun's gravitational pull (picture 3). As the young Sun became hotter, its solar wind blew away most of the remaining gas and dust in the Solar System. The planets close to the Sun were stripped of their atmospheres, and only four giant planets were able to hold on to their enormous bodies of gas (picture 4).

4. When the centre of the cloud reached 10 million °C (50 million °F) a nuclear reaction began, which turned hydrogen into helium. The force of this change blew away the loose gas and dust in the Solar System, leaving only the planets and larger pieces of debris.

Leftover material

Even after the planets were formed, a great deal of material remained, but it was no longer being thrown together with the force needed to create planets. The debris was instead left to orbit the Sun as rocky bodies called asteroids, icy chunks called comets, or smaller pieces of space dust known as meteoroids (picture 5).

5. Over millions of years, the Solar System settled down into the pattern we know today.

The Sun

The nearest star to Earth is more than just a bright light in the sky. It is what makes life on our planet possible. Billions of years in the future, it will also end life on Earth. The Sun is a yellow main sequence star, made up of helium and hydrogen. The huge amounts of heat and light energy that it gives off are the result of nuclear fusion in its core, where hydrogen is changed into helium.

Space visits

No human being will ever walk on the surface of the Sun. Apart from the fact that it is a ball of burning gas without a solid surface, the tremendous heat and its volatile nature make it impossible for manned spacecraft to approach too closely. However, in 1990 a solar probe called Ulysses was launched to look at the poles of the Sun. From Earth, we can only see the area around the Sun's equator.

- At any time there are over 100,000 short-lived spicules in the chromosphere. These straight jets of gas last up to ten minutes before appearing to melt into the corona.
- The energy from the core passes through the radiative zone, which is about 380,000 km (236,000 mi) thick.
- The core is a massive nuclear reactor, where hydrogen is fused to create helium.
- Solar flares are huge explosions taking place in the chromosphere. Each one is a million million times more powerful than the first nuclear bomb.

- Solar prominences are jets of flaming hydrogen that are held in the corona by the magnetic field of the Sun. They spurt from the Sun at enormous speeds and can reach 500,000 km (310,000 mi) in height. Some form an arc and are called looped prominences.
- The convective zone consists of circling currents, taking heat out towards the photosphere. It is about 140,000 km (87,000 mi) thick.
- The outer atmosphere of the Sun is called the corona. It is a halo of hot gas with a temperature of 1,000,000°C (1,800,000°F).
- Rising 1,000 km (620 mi) above the photosphere is the chromosphere, a reddish layer of hydrogen with a temperature that varies from 4,000 to 8,000°C (7,200 to 14,400°F).
- The white hot photosphere is made up of hydrogen at a temperature of 5,500°C (9,900°F).

- Darker areas are called sunspots. These areas of cooler gas happen when the Sun's magnetic field blocks the flow of heat from the core. Sunspot activity peaks every 11.5 years. Key dates are 1990, 2001 and 2012.

Facts and Figures

Diameter: 1,400,000 km (870,000 mi)

Age: 4.6 billion years

Distance from Earth: 149.6 million km (92.9 million mi)

Surface temperature: 5,500°C (9,900°F)

Core temperature: 15,000,000°C (27,000,000°F)

Mass: 332,946 x Earth

Luminosity: 390 billion billion megawatts

Future life: 5 billion years

The solar wind

Charged particles are constantly being given off by the Sun. They are known as the solar wind and are strongest when sunspot activity is at its height. When the solar wind reaches the Earth's magnetic field, the charged particles interact with gases in the Earth's atmosphere 10 km (6 mi) above the surface. The interaction causes the particles to emit light, which is seen from Earth as a bright white or multicoloured lightshow, most visible within the polar circles. In the northern hemisphere this is known as the *aurora borealis*. The southern hemisphere sees the *aurora australis*.

Eclipses

Ancient civilizations used to believe that a solar eclipse signalled impending disaster, and it is easy to see why. A total eclipse is an awe-inspiring sight. Birds and animals fall silent, stars and planets appear overhead, and for up to eight minutes the world below falls into night. As incredible as this sight may be, it is important never to look directly at the Sun, even during an eclipse.

Total eclipses

Solar eclipses occur when the Moon's orbit around Earth takes it in front of the Sun, blocking its light. When this happens, the Moon and the Sun appear to be exactly the same size in the sky. This is because although the Moon is much smaller, it is also much closer. During a total eclipse, the Moon covers up the Sun completely. All that can be seen of the Sun is its faint outer atmosphere, the corona, like a cloud of gas around a dark centre.

- **Total eclipses occur once every 18 months around our planet. However, it is estimated that any one place on Earth only sees a total eclipse every 360 years.**

WARNING!
Never look directly at the Sun, even during an eclipse. This can cause blindness. See page 113 for a safe method of observing the Sun.

Special effects

A total eclipse of the Sun can be a frightening sight, turning day into night in spectacular fashion. However, as the Moon appears to devour its celestial ruler, some incredibly beautiful "special effects" take place. Just before the Sun disappears, a brilliant bright spot can be seen on the edge of the Moon, like the diamond on a ring. This is caused by the last fingers of the Sun's light filtering through valleys and mountain ranges on the Moon. Sometimes the bright spot can appear as an arc of glowing pearls, an effect known as Bailey's Beads.

Facts and Figures

Solar eclipses:	
21 June 2001	Total
14 December 2001	Annular
10 June 2002	Annular
4 December 2002	Total
31 May 2003	Annular
23 November 2003	Total
8 April 2005	Total
3 October 2005	Annular
1 August 2008	Total
26 January 2009	Annular
2 July 2009	Total
11 July 2010	Total
13 November 2012	Total

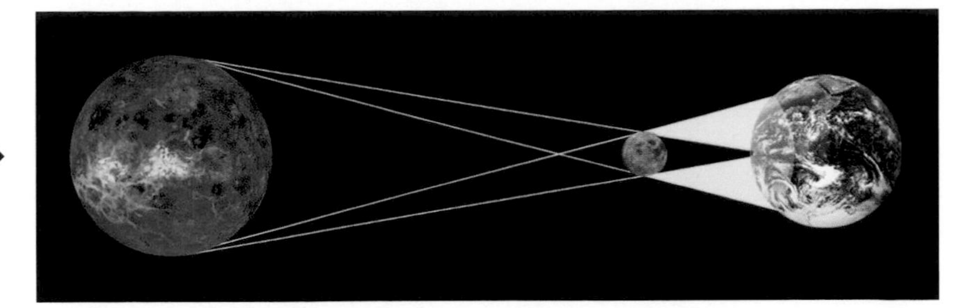

- **A solar eclipse occurs when the Moon comes between the Earth and the Sun. The Moon blocks out the Sun's light, and its shadow falls on Earth's surface. People at different locations on Earth will see different things. Anybody inside the complete shadow of the Moon, called the umbra, will see a total eclipse. Anybody in the penumbra, the lighter, outer shadow, will only see a partial eclipse.**

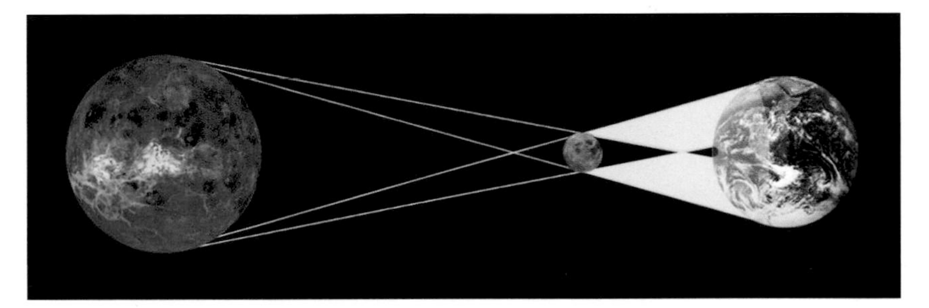

- **During an annular eclipse, the Moon is at its furthest point from Earth. Because of this, its shadow does not reach the Earth's surface. Those directly beneath the point of the umbra will see the Sun's photosphere like a ring of fire around the Moon.**

Annular eclipses

The Moon's orbit around Earth is elliptical rather than circular. This means that at some times the Moon is farther away from Earth than at others. Because of this, the Moon is sometimes not big enough to cover the Sun completely during an eclipse. This is called an annular eclipse because a ring of brilliant sunlight, which looks like a circle of fire, appears around the Moon.

- **This image is formed from several different photographs taken as the Moon covers up the Sun. This is an annular eclipse, as the Moon is not close enough to Earth to cover the Sun completely.**

The Planets

As far as scientists know, there are eight planets in our Solar System. Each was created from the same cloud of particles that gave the Sun its power, but they are all startlingly different in size and composition. Venus is a death-trap for anybody bold enough to step onto its boiling surface. Jupiter and Saturn are giant balls of gas with tiny solid cores. And Earth is positioned in exactly the right place for life to exist.

Facts and Figures

Time taken for the planets to form:

- Mercury: 80,000 years
- Venus: 40,000 years
- Earth: 110,000 years
- Mars: 200,000 years
- Jupiter: 1,000,000 years
- Saturn: 9,000,000 years
- Uranus: 300,000,000 years
- Neptune: 1,000,000,000 years (1 billion years)

How planets formed

The planets formed from the same swirling spiral of debris as the Sun. In this violent whirlpool of particles, tiny pieces of debris were thrown together to form larger rocky bodies called planetisimals. Gradually, these planetisimals joined together to form even larger objects called protoplanets. The protoplanets near the Sun joined to form the rocky planets: Mercury, Venus, Earth and Mars. Protoplanets that formed farther out grew large enough to attract vast quantities of gas and became the giant planets: Jupiter, Saturn, Uranus and Neptune. The dwarf planet Pluto is thought to have been formed from the leftover material.

Seen here (from top to bottom) are Mercury, Venus, Earth and its Moon, Mars, Jupiter, Saturn, Uranus, Neptune, and the dwarf planet Pluto.

The diagram below, which is not to scale, shows the average distances of the planets from the Sun in astronomical units (AU).

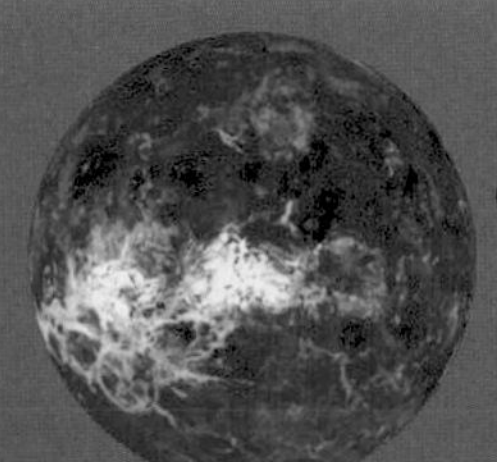

Mercury: 0.4 AU

Venus: 0.7 AU

Earth: 1 AU

Mars: 1.5 AU

Jupiter: 5.2 AU

Saturn: 9.5 AU

● Leftover material in the Solar System collided with the surfaces of the young planets. These impacts, billions of years ago, created the craters that we can still see today.

● Because of the huge distances involved in space, astronomers often measure in astronomical units. One astronomical unit is equal to the Earth's average distance from the Sun, approximately 149,600,000 (149.6 million) km (92,960,000 mi).

Planetary names

The names of the planets in our Solar System originate from characters in ancient Greek and Roman legends. Mercury was named after the nimble messenger of the gods because of its fast orbit around the Sun. Venus was named after the goddess of beauty and love because of its brightness in the night sky. Mars became identified with the Roman god of war because of its angry red appearance. Jupiter was given the name of the king of the gods because of its size. Saturn was the god of the harvest. Uranus was named after the father of the gods, and Neptune after the Roman god of water.

● Jupiter, king of the Roman gods, was believed to be enormously powerful.

The planets

The eight planets in our Solar System form two distinct groups. The inner planets are those closest to the Sun, consisting of Mercury, Venus, Earth and Mars. These are much smaller than the outer planets, and are made of rock and metal. Jupiter, Saturn, Uranus and Neptune are the outer planets, which are giant balls of gas with small, solid cores. Additionally, there is the dwarf planet known as Pluto, in the Kuiper Belt, which is much smaller than its gassy neighbours and is made of rock, metal and snow.

Uranus: 19.2 AU

Neptune: 30.1 AU

Pluto: 39.5 AU

Mercury

Mercury is the first planet in our Solar System, orbiting a mere 58 million kilometres (36 million miles) from the Sun. Despite this, temperatures on the planet can drop to almost –200°C (–330°F) at night. This is because Mercury is little more than a floating rock, with barely any atmosphere and no life under its scarred surface.

Hidden planet

Mercury is a very difficult planet to explore. It is normally obscured by the Sun's glare when viewed from Earth. Even the Hubble Space Telescope cannot observe the planet because the strength of the Sun's rays might damage its equipment. Only one probe, Mariner 10, has made it as far as Mercury, although its flight path meant that it could only photograph one side of the planet. The other half we know very little about.

Facts and Figures

- Diameter: 4,877 km (3,031 mi)
- Distance from Sun: 58 million km (36 million mi)
- Average surface temperature: 170°C (340°F)
- Surface gravity: 0.38 x Earth
- Length of orbit: 88 Earth days
- Length of day: 1,408 Earth hours
- Mass: 0.055 x Earth
- Density (water = 1): 5.42
- Number of moons: 0
- Number of rings: 0

Craters are formed by space rocks, called meteorites, hitting the surface of a planet or moon.

When a rock crashes into a planet like Mercury, it creates an enormous dent in the ground.

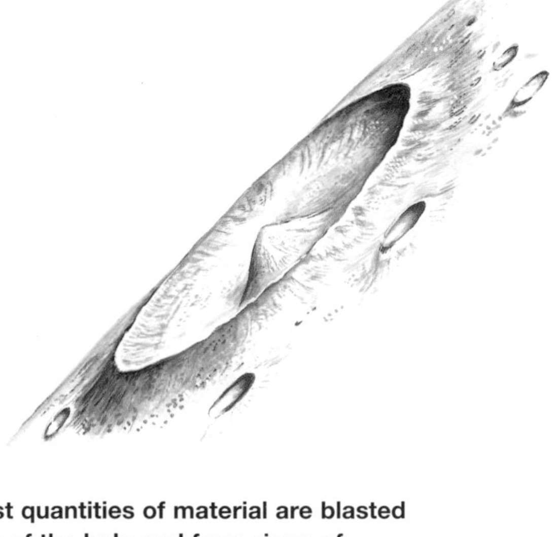

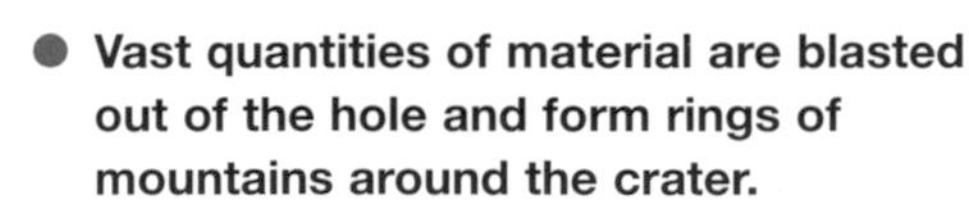

Vast quantities of material are blasted out of the hole and form rings of mountains around the crater.

Mercury's orbit

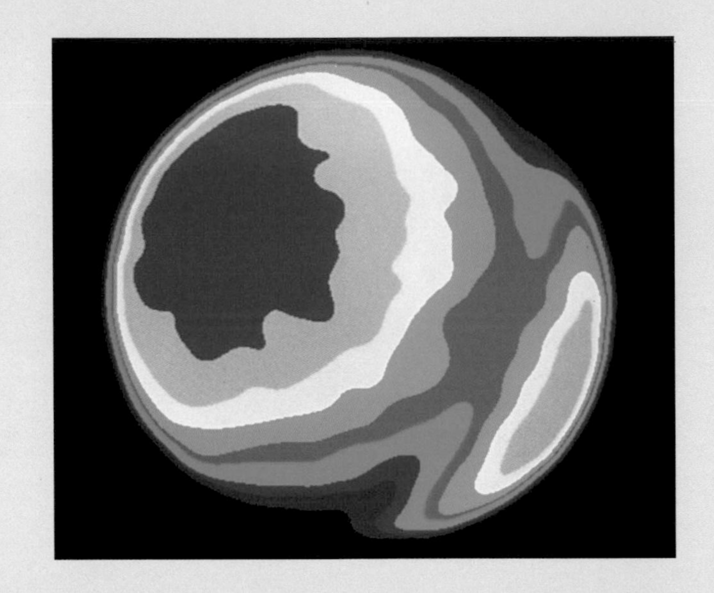

Mercury's orbit close to the Sun means that it has the greatest variation between day and night temperature of any planet in the Solar System. During the day the surface temperature can reach 450°C (840°F). However, the lack of atmosphere on the planet means that heat cannot be retained at night, and the temperature can drop by over 600°C (1,100°F). Parts of Mercury become so cold that some scientists believe there is ice on the planet. Meteorites may have carried water to Mercury's surface, and ice may have formed in deep, shadowed craters that never see the Sun.

- **Mercury takes 59 days to spin once on its axis. However, by the time it has spun once it has also travelled two-thirds of its 88-day orbit around the Sun. Because of this, the time from one midday to the next on Mercury is 176 days. A person born on Mercury would be 2 years older each day!**
- **Some meteorite impacts have been so powerful that they have sent shockwaves through the planet and created mountain ranges on the other side.**
- **Radar imaging of the planet has revealed areas of high reflectivity near the planet's poles. Scientists believe this may be ice.**

A dead planet

All planets are born from countless collisions of rocks and space debris. They are formed with the heat of these collisions locked deep within their cores. A planet is "alive" as long as it retains this heat within it. We can tell whether this is the case by looking for volcanic activity. Mercury, like our Moon, has a heavily cratered face, which shows that there has been no volcanic activity on the planet for billions of years. This makes Mercury a "dead" planet.

Life on Earth

Our planet is remarkable because it contains something that scientists have found nowhere else in space – life. Earth is home to an incredible variety of life forms, from single cell bacteria to complex organisms such as human beings. There are millions of different species of living creatures on Earth, but the origins of life on our planet are unclear.

Why Earth?

Conditions on Earth are perfect for life to take hold. Earth is near the middle of the ecosphere, the zone around the Sun where it is neither too hot nor too cold for life. As a result, ours is the only planet in the Solar System to have liquid water on its surface. Water is a vital ingredient for life, and every living creature on the planet uses it to survive—from frozen algae living in rocks in the middle of the Antarctic to plants that live in the driest deserts.

Origins of life

Nobody knows exactly what it takes for life to begin, but scientists have considered several possibilities. Some suggest that living cells were carried to Earth by comets. Halley's Comet was investigated by the Giotto probe in 1986, and its nucleus was found to contain molecules that were similar to living cells. If a comet such as this collided with Earth at just the right time, life may have taken hold. Others believe that lightning bolts that flashed through Earth's early atmosphere caused chemical reactions in the air. These changes led to living cells being created.

4. Fish developed and evolved into amphibians. As the Earth became drier, amphibians evolved into reptiles that could survive better on land.

7. Today, we are realizing that destructive human behaviour may pose a threat to all life.

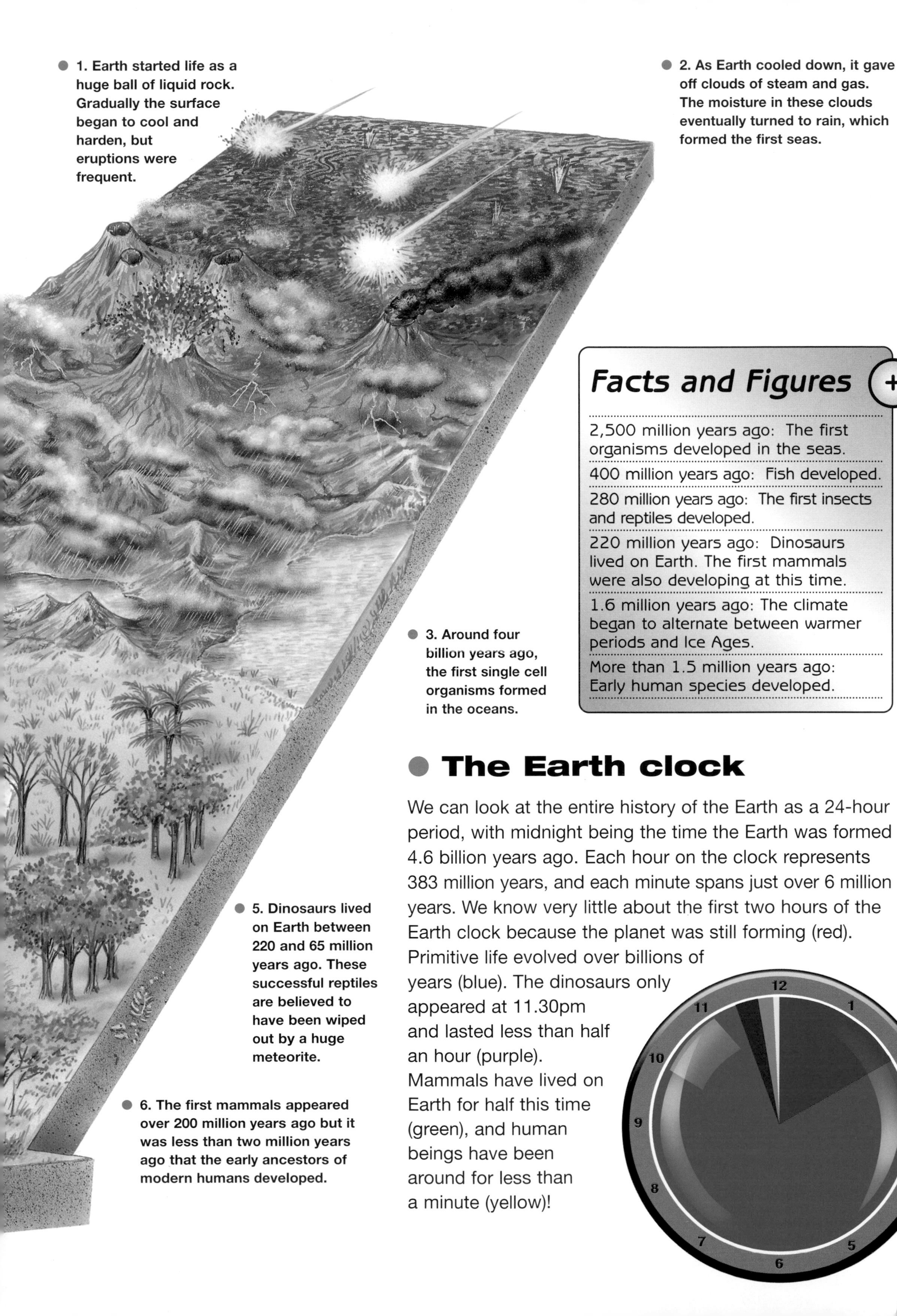

1. Earth started life as a huge ball of liquid rock. Gradually the surface began to cool and harden, but eruptions were frequent.

2. As Earth cooled down, it gave off clouds of steam and gas. The moisture in these clouds eventually turned to rain, which formed the first seas.

3. Around four billion years ago, the first single cell organisms formed in the oceans.

5. Dinosaurs lived on Earth between 220 and 65 million years ago. These successful reptiles are believed to have been wiped out by a huge meteorite.

6. The first mammals appeared over 200 million years ago but it was less than two million years ago that the early ancestors of modern humans developed.

Facts and Figures

2,500 million years ago: The first organisms developed in the seas.

400 million years ago: Fish developed.

280 million years ago: The first insects and reptiles developed.

220 million years ago: Dinosaurs lived on Earth. The first mammals were also developing at this time.

1.6 million years ago: The climate began to alternate between warmer periods and Ice Ages.

More than 1.5 million years ago: Early human species developed.

The Earth clock

We can look at the entire history of the Earth as a 24-hour period, with midnight being the time the Earth was formed 4.6 billion years ago. Each hour on the clock represents 383 million years, and each minute spans just over 6 million years. We know very little about the first two hours of the Earth clock because the planet was still forming (red). Primitive life evolved over billions of years (blue). The dinosaurs only appeared at 11.30pm and lasted less than half an hour (purple). Mammals have lived on Earth for half this time (green), and human beings have been around for less than a minute (yellow)!

The Moon

The Moon is Earth's nearest neighbour in space. It orbits our planet at a distance of 384,400 km (238,850 mi) and always keeps the same side pointing towards us. It may seem a long way away, but the Moon is close enough to have a tremendous effect on conditions on earth. It is the pull of the Moon's gravity that causes the tides in the seas.

Facts and Figures

- Diameter: 3,476 km (2,160 miles)
- Age: 4.6 billion years
- Distance from Earth: 384,400 km (238,000 miles)
- Surface temperature: from –155°C to 105°C (–247°F to 221°F)
- Time taken to orbit the Earth: 27.3 days
- Mass: 0.012 x Earth
- Time taken to turn on its axis: 27.3 days

- **Craters on the near side of the Moon have been named after famous people, such as astronomers.**
- **Although space exploration gave us information for the first time about the side of the Moon that is always turned away from Earth (the "dark side"), there are still areas of the Moon that have not been mapped, particularly around the south pole.**

The Moon's surface

The Moon has no light of its own, but like Earth is lit by light from the Sun. Even with the naked eye, some features of the Moon's surface can be seen. In fact, it is a dry, barren landscape, pitted with craters caused by the bombardment of meteorites over three billion years ago. The Moon has no rain, wind or earthquakes to wear away or break down the craters, so they have remained the same for millions of years.

Lunar influences

The Moon is a mystery because we do not know exactly where it came from. Before scientists began examining lunar rocks, they had several theories about how the Earth acquired its satellite. Some believed that when the Earth was young, it was spinning at such a high speed that it lost some of its molten form. This globule of lava solidified and became the Moon. The most popular theory today is that when Earth was young, it was hit by a Mars-sized object. This immense collision sent a large amount of material into space. The material joined to form the Moon.

Some craters are called "ray craters". They were formed when meteorites struck the Moon about 800 million years ago, hurling bright rocks into a pattern of "rays" around the point of impact.

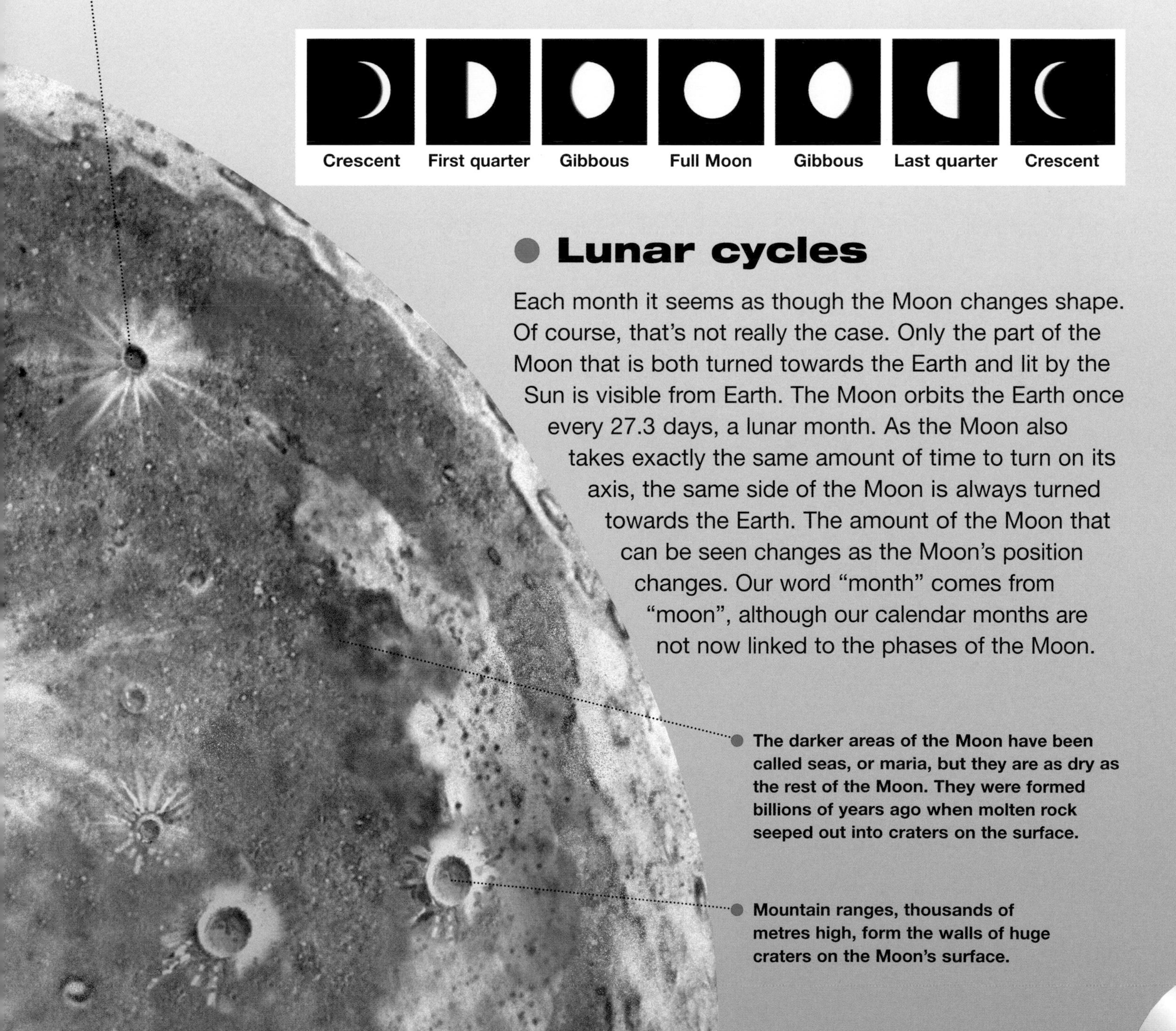

Crescent | First quarter | Gibbous | Full Moon | Gibbous | Last quarter | Crescent

Lunar cycles

Each month it seems as though the Moon changes shape. Of course, that's not really the case. Only the part of the Moon that is both turned towards the Earth and lit by the Sun is visible from Earth. The Moon orbits the Earth once every 27.3 days, a lunar month. As the Moon also takes exactly the same amount of time to turn on its axis, the same side of the Moon is always turned towards the Earth. The amount of the Moon that can be seen changes as the Moon's position changes. Our word "month" comes from "moon", although our calendar months are not now linked to the phases of the Moon.

The darker areas of the Moon have been called seas, or maria, but they are as dry as the rest of the Moon. They were formed billions of years ago when molten rock seeped out into craters on the surface.

Mountain ranges, thousands of metres high, form the walls of huge craters on the Moon's surface.

Mars

Mars is often called the red planet. It has a barren and hostile appearance. However, it is the planet on which scientists are most optimistic about finding life. Whilst the discovery of little green men is doubtful, there is evidence to show that, billions of years ago, the atmosphere on Mars could have sustained primitive life. Some even believe that it still exists there today.

Facts and Figures

- Diameter: 6,785 km (4,216 mi)
- Distance from Sun: 228 million km (142 million mi)
- Surface temperature: –63°C (–81°F)
- Surface gravity: 0.38 x Earth
- Length of orbit: 687 Earth days
- Length of day: 25 Earth hours
- Mass: 0.107 x Earth
- Density (water = 1): 3.94
- Number of moons: 2
- Number of rings: 0

- **Mars has many channels on its surface, which scientists believe were formed by rapidly moving water.**

- **In 1997, the Pathfinder spacecraft landed on Mars to explore the planet. It carried a six-wheeled rover called Sojourner to investigate rock and soil samples (shown left).**

Water on Mars

The red planet does not have water on its surface now, but there is evidence to show that there were rivers and seas on Mars in the past. The Viking probes took many pictures of the planet's surface, and scientists have identified several channels that could only have been formed by running water. Astronomers believe that there was a great deal of liquid water on Mars billions of years ago. Meteors and volcanic activity kept the planet warm, so the water did not freeze. Nowadays, however, Mars is too cold for water to exist in liquid form, and scientists believe that if it exists, it is as ice under the surface.

Little green men

Primitive life, such as single cell organisms, may have existed in the watery channels on Mars billions of years ago.

The belief in life on Mars has been around for centuries. As far back as the nineteenth century, an astronomer named Schiaparelli reported seeing channels on the surface of Mars. Many believed that these had been created by intelligent life forms. Scientists now think the channels were created by natural forces, but claim that primitive life may have existed on Mars billions of years ago. A Martian meteorite that landed on Earth thousands of years ago contains microscopic structures that some believe may have been formed by living organisms.

This is a magnified image of the Martian meteorite ALH84001.

Surface of Mars

Mars has one of the most striking surfaces of any planet in the Solar System. Giant volcanoes tower above the landscape. The largest of these is Olympus Mons, which is 25 km (15.5 mi) tall, three times larger than Mount Everest! As well as this, Mars is home to Valles Marineris, a canyon that is 180 km (112 mi) wide, up to 7 km (4.3 mi) deep and long enough to stretch right across the United States of America.

Saturn

Saturn's elaborate network of rings makes it one of the most beautiful sights in the Solar System. However, scientists believe that these rings are only a recent addition to the planet, forming in the last few million years. Some even say that Saturn will not keep its rings for more than a few million years before they either disperse or form a new moon.

- Saturn is 95 times larger than Earth. However, it has the lowest density of all the planets. Saturn is less dense than water, so if it was placed in an enormous ocean, it would float!

- Saturn's rings are made up of millions of tiny pieces of debris. These chunks are made of rock and ice.

Saturn's rings

Saturn is the most distinctive planet in the Solar System because of its rings. What Galileo once described as the planet's "ears" have now been revealed as a set of fantastically complex rings, each made up of thousands of bands of debris. Nobody knows for sure how Saturn got its rings. Scientists believe they are all that remains of a moon that was wrenched apart by the gravity of Saturn. Some even believe that Saturn's gravity will cause the rings to group together and eventually reform a moon. The image above shows what Saturn's rings would look like from within.

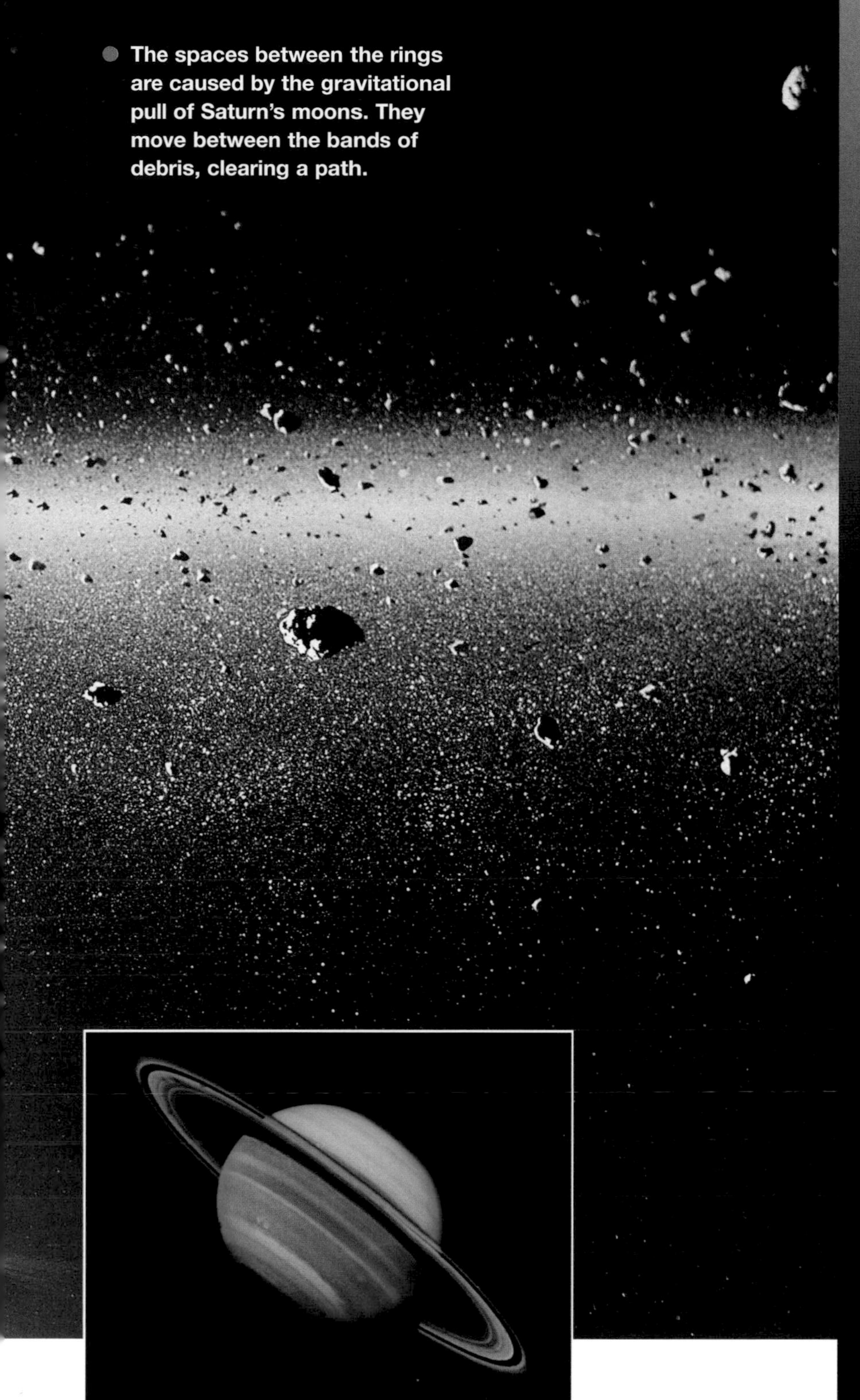

The spaces between the rings are caused by the gravitational pull of Saturn's moons. They move between the bands of debris, clearing a path.

Saturn's weather

Saturn may be thought by many to be a more beautiful planet than its neighbour Jupiter, but it is no less violent. Like Jupiter, it spins so fast that it bulges at its centre, causing ferocious winds. It is made from the same material as Jupiter, and every 30 years a raging storm breaks out on Saturn and spreads across the entire planet. The bands of colour on the image above show weather patterns developing on Saturn.

Facts and Figures

Diameter: 120,536 km (74,900 mi)
Distance from Sun: 1,427 million km (886 million mi)
Surface temperature: –180°C (–290°F)
Surface gravity: 0.92 x Earth
Length of orbit: 10,760 Earth days
Length of day: 10 Earth hours
Mass: 95.2 x Earth
Density (water = 1): 0.69
Number of moons: 18
Number of rings: 7

Saturn's moons

Saturn has the record for the most moons in the Solar System. Astronomers know of the existence of 18, but there may be more. Titan is the only moon in our Solar System that has a substantial atmosphere. Its clouds are too thick to see through, but scientists believe the conditions on the moon's surface are similar to those on Earth billions of years ago. In 2004 the Huygens probe landed on Titan, as shown in the picture below, in order to complete experiments that may help scientists discover the origins of life on Earth.

Uranus

Uranus is named after the Greek father of the gods, but many question whether it deserves such a grand title. This is because Uranus is thought by many to be the blandest planet in the Solar System. However, Uranus has many interesting features of its own. It rolls around the Sun like a barrel, and its weather, although it cannot be seen by the naked eye, is highly active.

The infrared image above shows bands of violent weather activity on the cloud tops of Uranus.

The barrel planet

Uranus is tipped on its side and is the only planet to roll around the Sun like a barrel instead of spinning upright. Its 11 faint, black rings and 17 moons spin around it like a ferris wheel. It is thought that this is because an object the size of Earth collided with Uranus billions of years ago. The incredible force of the impact knocked the planet onto its side. This means that each pole has 21 years of continual sunlight, then is plunged into 21 years of perpetual night.

Uranus has very faint, black rings that were discovered by chance in 1977. Astronomers saw that as Uranus moved through the sky, the light from stars behind it twinkled. Scientists realized that the planet's rings were causing this to happen.

Hidden weather

Uranus is often unfairly labelled the most boring planet in the Solar System. A thick layer of methane in the upper clouds absorbs red light, giving the planet its uniform aquamarine colour. There is little evidence on Uranus' bland face of much activity. Viewed normally, the planet has the complexion of a snooker ball. However, the Voyager space probe's infrared pictures of the planet showed it to have bands of weather activity just like the planets Jupiter and Saturn.

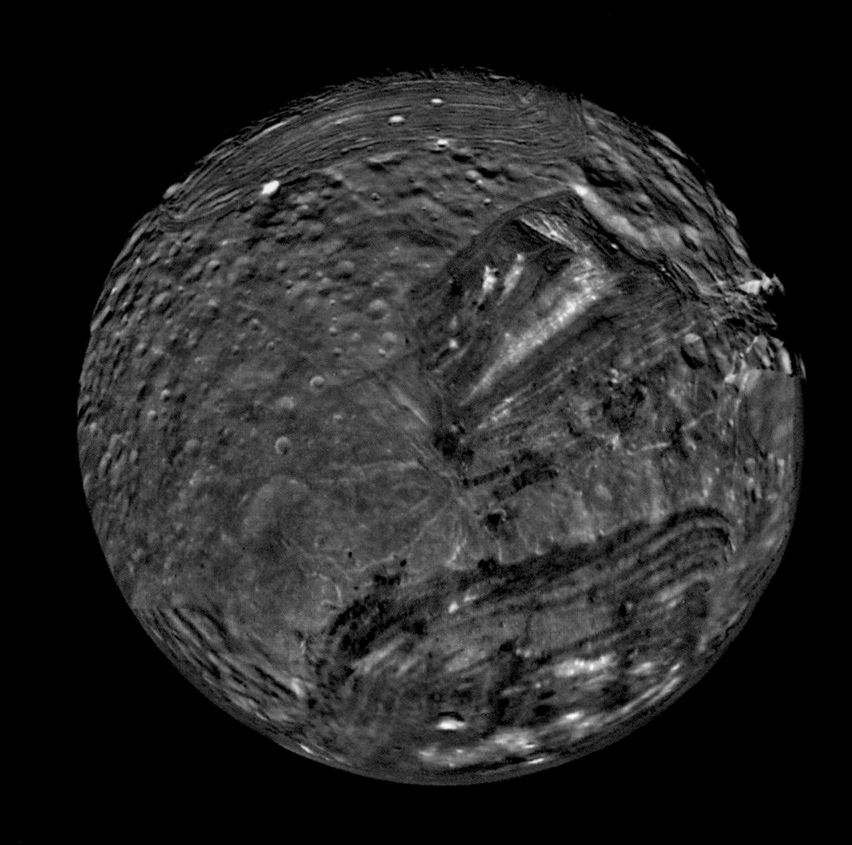

Facts and Figures

- Diameter: 51,166 km (31,794 mi)
- Distance from Sun: 2,869 million km (1,783 million mi)
- Surface temperature: –195°C (–319°F)
- Surface gravity: 0.89 x Earth
- Length of orbit: 30,065 Earth days
- Length of day: 18 Earth hours
- Mass: 14.5 x Earth
- Density (water = 1): 1.27
- Number of moons: 17
- Number of rings: 11

Chaotic faces

Uranus itself may be devoid of surface features, but its moons carry a fascinating portrait of a violent past. The bizarre alien landscapes of many of Uranus' moons are thought to have been created by water. Scientists believe that as liquid water rose from the hot interior of the moons, it froze and expanded. This caused the crust to buckle outwards. Miranda has the most chaotic surface of any moon. Some parts of her core are now on her surface, and some parts of her crust have been buried. This is most likely to be because the moon has at one time been pulled apart and has gradually reformed.

Some scientists believe that planets get their rings when a moon is pulled apart by conflicting gravitational forces.

The moon will be pulled one way by its orbit and the opposite way by the gravity of its parent planet, causing it to split and crumble.

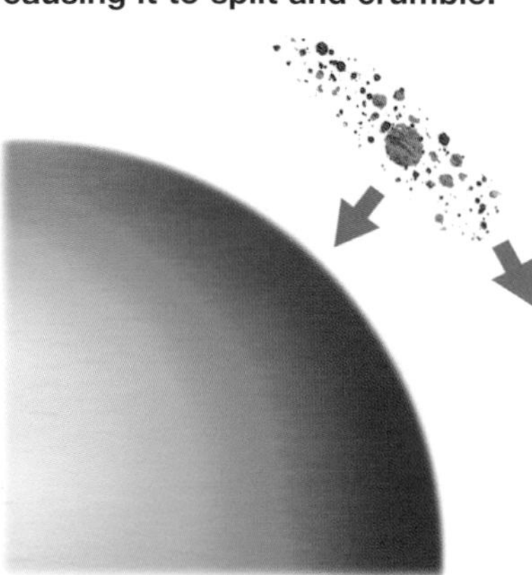

Neptune

Neptune is the outermost of the four gas giants and is in many ways Uranus' sister planet. They are almost identical in size and composition, and even affect one another's orbits around the Sun. However, Neptune is a far more exciting planet, with violent weather systems that make hurricanes on Earth seem insignificant. Nobody born on Neptune would live to be one year old, as the planet has a 165-year orbit.

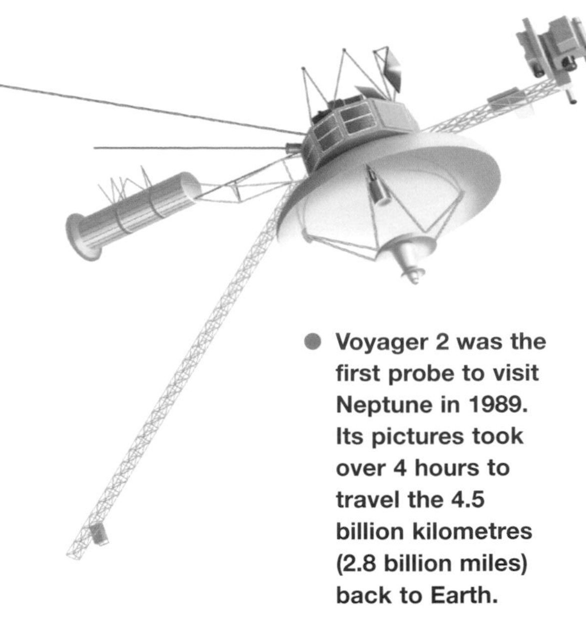

Voyager 2 was the first probe to visit Neptune in 1989. Its pictures took over 4 hours to travel the 4.5 billion kilometres (2.8 billion miles) back to Earth.

Windy planet

Neptune shares Uranus' bland appearance. However, unlike its sister planet, Neptune boasts one of the most violent weather systems in the Solar System. Storms the size of Earth rage continuously in the planet's atmosphere. Neptune's face is always changing as these storms come and go. Winds of up to 2,000 km/h (1,240 mph) can be found here, the fastest on any planet. These violent conditions are caused by the heat inside the planet. Temperatures in the planet range from –220°C (–364°F) on the cloud tops to an incredible 7,000°C (12,630°F) in the planet's core. This is hotter than the surface of the Sun.

Neptune has a dark spot just like Jupiter (shown left). This is a raging storm as large as Earth!

Like its gassy neighbours, Neptune has a series of five very faint rings. They are so dark that only Voyager 2's cameras could see them.

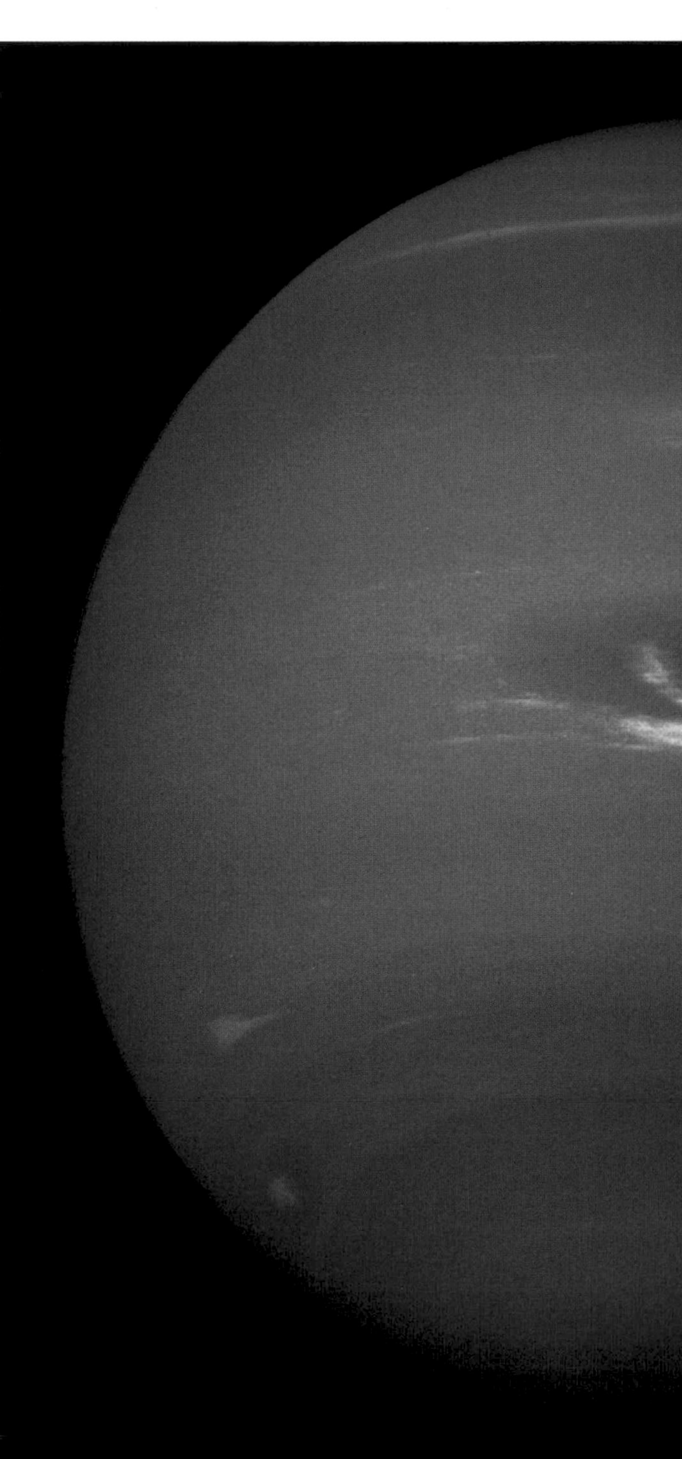

Facts and Figures

- Diameter: 49,557 km (30,795 mi)
- Distance from Sun: 4,496 million km (2794 million mi)
- Surface temperature: –200°C (–330°F)
- Surface gravity: 1.13 x Earth
- Length of orbit: 60,190 Earth days
- Length of day: 19 Earth hours
- Mass: 17.1 x Earth
- Density (water = 1): 1.71
- Number of moons: 8
- Number of rings: 6

Triton

One of Neptune's most fascinating features is actually one of its moons, Triton. This has been found to be the coldest world in the Solar System. Its surface can reach temperatures as low as –235°C (–391°F). This is a mere 38°C (100°F) above the coldest temperature possible, *absolute zero.* Not only this, but Triton is also geologically active, spewing volcanoes of nitrogen gas deep into its atmosphere. The moon is unusual, too, in that it orbits in the opposite direction to its mother planet.

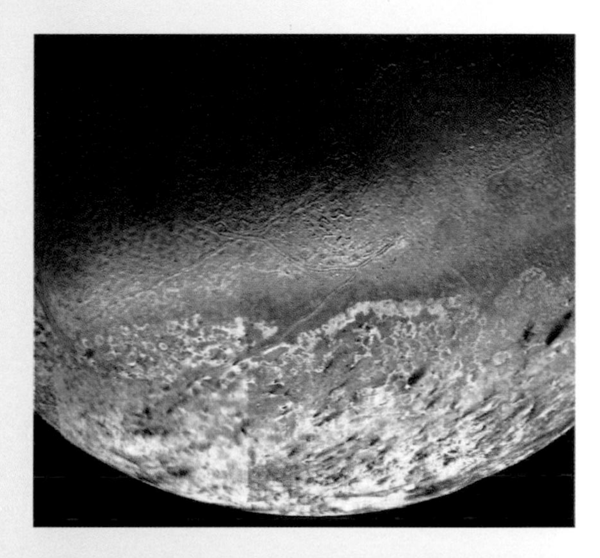

Triton is Neptune's largest moon. It is so enormous that it is bigger than the planet Pluto. Scientists believe that it may once have been a separate body that was caught by Neptune's gravitational pull.

Neptune's atmosphere contains much more of the gas methane than Uranus'. This is what makes Neptune bluer. Bright clouds of methane appear as white streaks.

The ice giants

The discovery of Neptune forced scientists to rethink their ideas about the gas giants. They discovered that Uranus and Neptune are very different to Jupiter and Saturn, describing the outer two as ice giants rather than gas giants. They are much younger than their bigger, gassy neighbours, and so were not able to draw on the vast clouds of hydrogen and helium that made Jupiter and Saturn so enormous. Beneath their cloud tops, Uranus and Neptune may have oceans of water, heated by their solid cores.

Pluto

Pluto is a dwarf planet, smaller than our moon, in the Kuiper Belt. At nearly 5,900 million kilometres (3,670 million miles) from the Sun, it has never been visited by a probe. Discovered in 1930, Pluto was considered to be the ninth planet of the solar system, until it was re-classified by the International Astronomical Union in 2006.

- Pluto takes 248 years to orbit the Sun. Amazingly, this means that not even half a year has elapsed on Pluto since its discovery in 1930!

- Pluto is different from the inner planets, which are much denser and composed of large amounts of iron and nickel, and it is vastly different from the gas giants.

- Pluto's strange orbit means that for periods of around 20 years it moves closer to the Sun than Neptune. It was like this between 1979 and 1999.

Twin planets

In 1978, Pluto was found to have a companion moon, which was named Charon. The moon is one third the size of Pluto, and the two bodies are a mere 20,000 kilometres (12,430 miles) apart, and are caught in a gravitational headlock, forming what scientists call a "dual planet system". Nobody knows how Pluto managed to adopt so large a moon. Some believe Charon is made from ice chipped off Pluto by a collision, others that the worlds formed in different places and were somehow caught by one another later on.

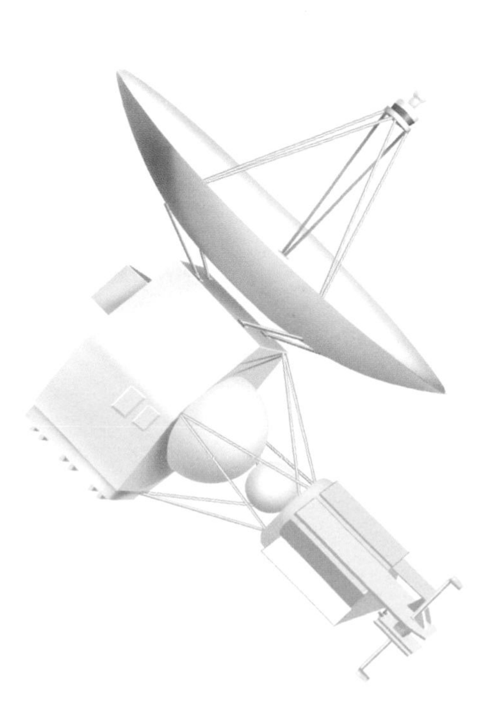

- **Pluto's orbit is more elongated than any of the eight planets in the Solar System. It is tilted against the plane of its neighbours, so that it spends half its year above the eight planets, then half its year below.**

Facts and Figures

Diameter: 2,274 km (1,413 mi)
Distance from Sun: 5,899 million km (3,666 million mi)
Surface temperature: –230°C (–380°F)
Surface gravity: 0.067 x Earth
Length of orbit: 90,800 Earth days
Length of day: 153 Earth hours
Mass: 0.002 x Earth
Density (water = 1): 2.03
Number of moons: 1
Number of rings: 0

Planet or not?

Many scientists questioned Pluto's former status as a planet, arguing that it was too small and its orbit too elongated to be classified as a proper planet. Some scientists claimed that Pluto was the largest of the objects in the Kuiper Belt, a collection of rocky and icy debris orbiting beyond Neptune. Others argued that Pluto and Charon were actually moons of Neptune that broke free billions of years ago and were caught in their own orbit around the Sun. In 2006, as the result of this debate, the International Astronomical Union re-classified Pluto as a dwarf planet in the Kuiper Belt.

A dark planet

In summer, Pluto has a slight atmosphere because the surface warms up enough to melt some of its ice, turning it to gas. As Pluto moves away from the Sun, the gas freezes and becomes ice again. This means that in winter, Pluto's weather doesn't just become worse, it completely disappears! On Pluto it is always dark and cold, even in the middle of the day. This is because the Sun appears 1,000 times fainter from the surface of Pluto than it does from Earth, little more than a faint star.

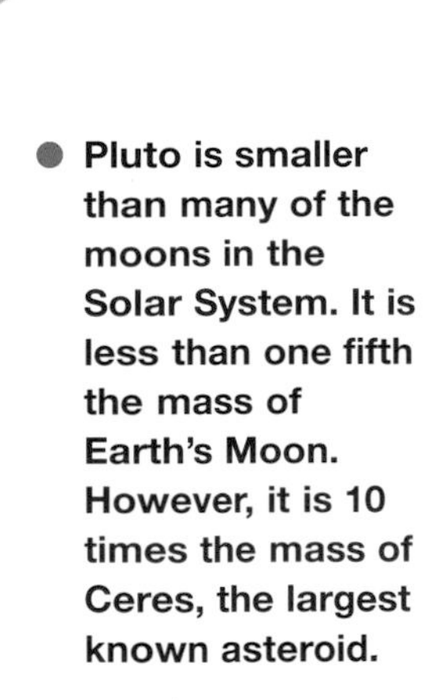

- **Pluto is smaller than many of the moons in the Solar System. It is less than one fifth the mass of Earth's Moon. However, it is 10 times the mass of Ceres, the largest known asteroid.**

Comets

Many hundreds of years ago, the sight of a comet blazing across the sky was feared by millions worldwide. It was believed that it was a bad omen, foretelling danger in the near future. Nowadays, comets are thought by many to be the most spectacular of all astronomical sights, and scientists believe that they may contain the key to the origins of life on Earth.

What is a comet?

A comet is very much like a dirty snowball. Its nucleus is made from dirty ice, frozen gas and rocky dust. These are relics from the birth of the Solar System 4.6 billion years ago. Most comets are just rocky lumps that orbit the Sun from a distance. However, when a comet's orbit takes it closer to the Sun, it begins to change. The Sun's heat warms the ice, and as the nucleus melts, gas and dust are released into a cloud called the coma. The Sun's solar wind blows the coma away from the nucleus, forming a tail that can stretch for hundreds of millions of kilometres.

Facts and Figures

Periods between comet appearances:	
Encke's Comet:	3.3 years
Pons–Winnecke	6 years
Giacobini–Zinner	6.5 years
Holme's Comet	6.9 years
Schwassmann–Wachman	16.2 years
Olber's Comet	74 years
Halley's Comet	76.3 years
Comet Kohoutek	75,000 years
Great Comet of 1864	2,800,000 years

Orbits

In the 1950s, an astronomer named Jan Oort claimed that comets originate from an enormous cloud of icy debris located far beyond Pluto. Pieces of what is now called the Oort Cloud can be dislodged and are sent on immense, elliptical orbits of the Sun. Sometimes comets can be pulled off their paths by the gravitational force of a planet or the Sun. In 1994, the Shoemaker–Levy 9 comet was pulled into Jupiter. It impacted at more than 200,000 km/h (124,000 mph), creating balls of fire larger than Earth!

The nucleus is the only solid part of a comet. It is made up of ice, dust and rock. These are all materials left over from the formation of the Solar System 4.6 billion years ago.

- A comet usually has two tails. A blue-coloured gas tail and a yellow or white dust tail. The dust tail is usually curved because of the Sun's gravity.

- A comet's tails can stretch for hundreds of millions of kilometres into space. They will always point away from the Sun.

- Scientists believe that comets are rich in organic material. Some even think it was a comet that brought life to Earth.

- The coma is formed from the gas released by ice as it melts. It can measure up to 100,000 km (62,000 mi) across!

Halley's Comet

Halley's Comet is perhaps the most famous comet of all. It can be seen from Earth every 76 years and was even recorded in 1066 on the Bayeux tapestry. Because scientists could predict when the comet would fly past Earth again, they were able to send a probe to investigate it. In 1986, the European space probe Giotto flew into Halley's Comet and photographed its nucleus in incredible detail, proving that comets are made of ice. Scientists are interested in finding out more about comets because they contain material from the origin of the Solar System.

- The Giotto probe, shown left, was able to gather data from inside Halley's Comet for ten hours. It photographed the nucleus of the comet, producing amazing images like the one shown above.

- Comets that take less than 200 years to orbit the Sun are called short-period comets. Long-period comets may take millions of years to complete one orbit!

Meteoroids

The Solar System is teeming with much smaller objects than comets or asteroids. These pieces of debris are called meteoroids and at first sight seem to be little more than lifeless lumps of rock. However, when meteoroids enter Earth's atmosphere, they become something much more beautiful—shooting stars. Many fall to the ground. If you find one, you can have your very own piece of space!

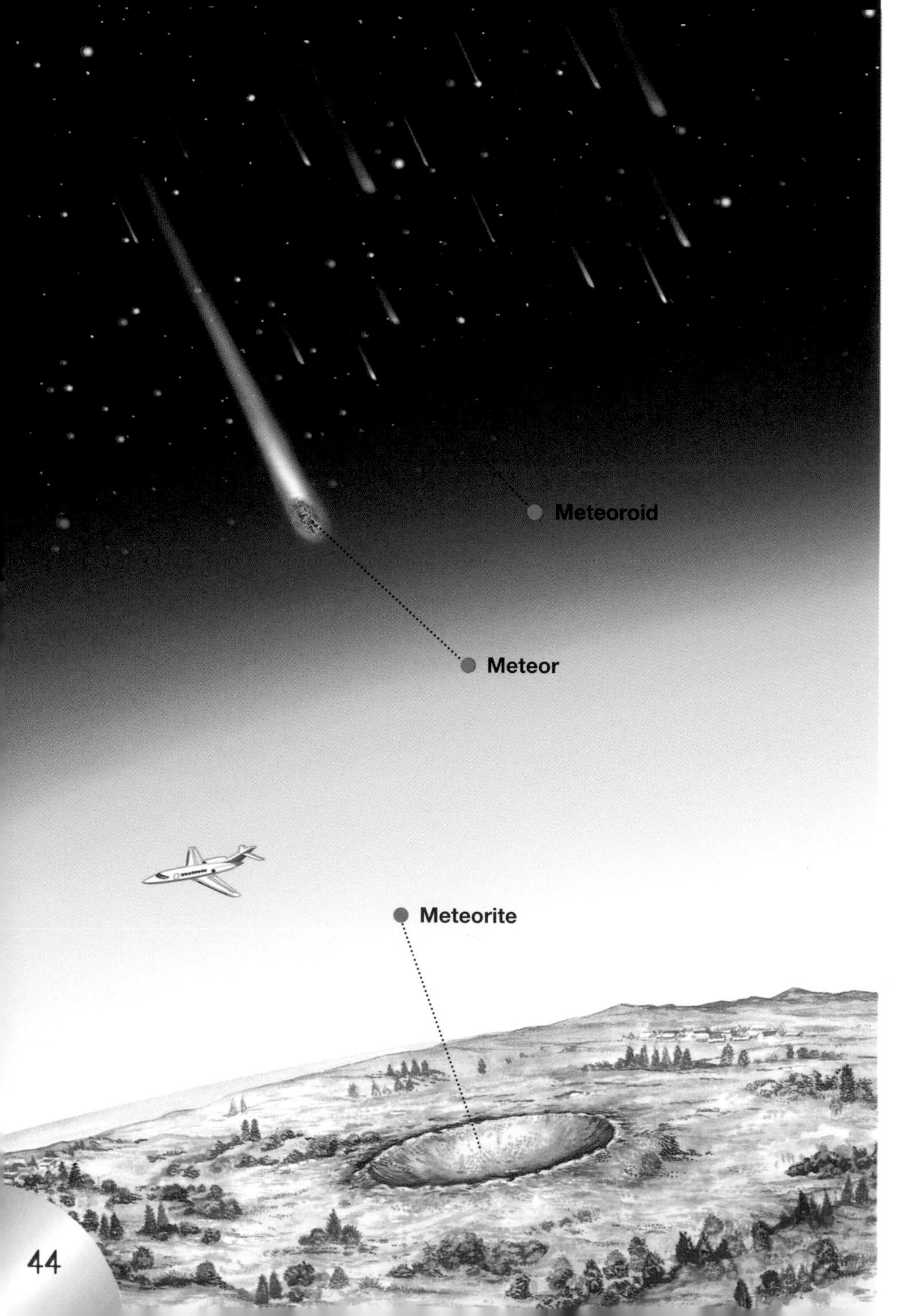

Facts and Figures

Yearly meteor showers:

Quadrantids: Boötes Constellation
1–6 January

April Lyrids: Lyra Constellation
19–24 April

Delta Aquarids: Aquarius Constellation
15 July – 15 August

Perseids: Perseus Constellation
25 July – 18 August

Taurids: Taurus Constellation
20 October – 30 November

Leonids: Leo Constellation
7–15 December

Which is which?

Meteoroids are made of rock or dust. They are mainly fragments of asteroids or comets that have been chipped or blown apart by a collision. These meteoroids travel through space and occasionally fly into Earth's atmosphere. When a meteoroid enters Earth's atmosphere, it burns up and begins to glow, becoming a meteor, better known as a shooting star. Most meteors burn up in the atmosphere before they hit the ground. Those that survive the trip and make it to Earth's surface intact are called meteorites. Meteorites are usually made from rock and iron, like the one shown below.

Meteors

Many meteoroids are formed from material lost by comets as they travel close to the Sun. A comet will lose material because of the force of the solar wind. These pieces of rock and dust are left in the comet's orbit, accumulating over many hundreds of years to form a trail of debris. If Earth passes through one of these trails, a shower of meteors appears. In fact, dozens of shooting stars can be seen every night around our planet. However, when Earth passes through a meteoroid stream, thousands can be seen in a short space of time.

Meteor showers can be predicted quite accurately because Earth passes through the meteor streams at roughly the same time each year.

Meteorites

Each year Earth puts on nearly 10,000 tonnes in weight due to meteoroids entering the atmosphere. When a meteoroid is too large to burn up in the atmosphere, it collides with the Earth's surface. Around 3,000 rocks fall onto Earth each year. Most fall into the sea and will never be found, but some crash into gardens and houses. In 1992, an empty car was flattened by a falling meteorite, but thankfully the chances of being hit by one are very remote!

Large meteorites leave enormous craters in the ground. The best preserved of these is the Barringer Crater in Arizona, USA (below), which was probably formed 30,000 years ago.

Scientists have discovered that many meteorites have come from the Moon and Mars. These may have been thrust into space by enormous collisions.

Our own Solar System may seem incredibly large and complex but it is only one of countless billions in the whole of space. Our Sun is one of over 200 billion stars in our galaxy alone, and there may be just as many galaxies in the depths of the Universe, each with a similar number of stars! Space is full of fascinating sights, from the beauty of newborn stars to the awesome power of black holes, which suck in everything around them.

We now know that numerous stars outside our Solar System have planetary systems of their own, and many astronomers believe that it is only a matter of time before extraterrestrial life is discovered.

Beyond the Solar System

Other Solar Systems

Centuries ago, it was considered heresy to suggest that the Earth was not the centre of the Universe, let alone that there may be other Earths in space! Nowadays, astronomers know for sure that there are planets orbiting nearby stars, although none have shown any sign of life. However, scientists are now hopeful that somewhere in the countless planetary systems in the Universe, Earth has a twin.

Facts and Figures

First extrasolar planets discovered:

Parent star:	Mass of planet:
51 Pegasi	150 x Earth
55 Cancri	270 x Earth
47 Ursae Majoris	890 x Earth
Tau Boötis	1,230 x Earth
Upsilon Andromedae	220 x Earth
70 Virginis	2,100 x Earth
16 Cygni B	480 x Earth
Rho Coronae Borealis	350 x Earth
Gliese 876	670 x Earth

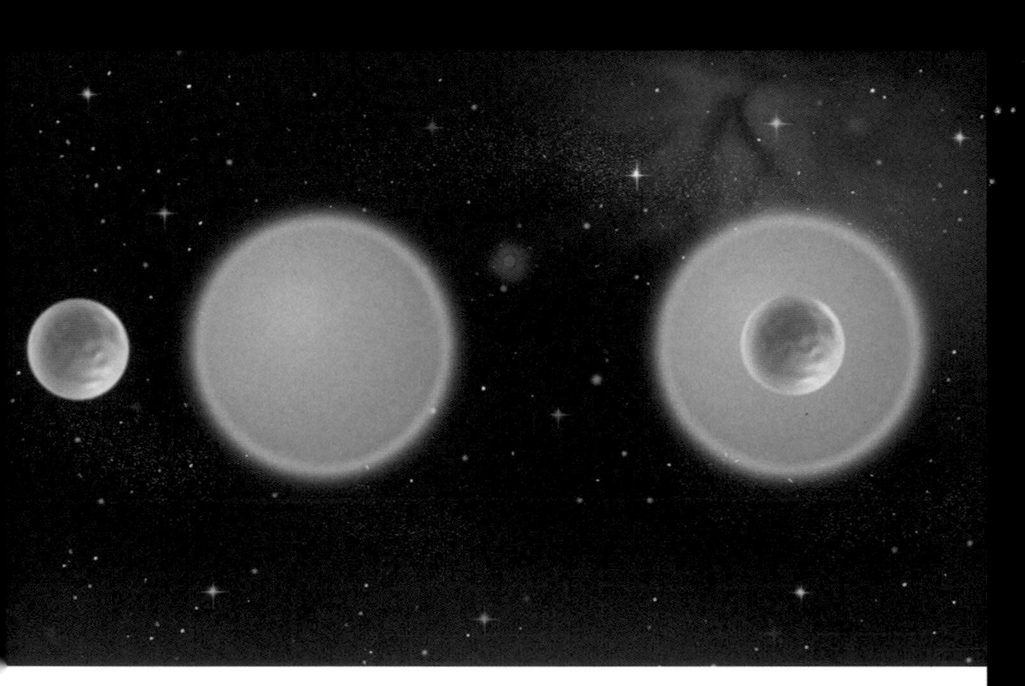

Most planets that astronomers have found so far are as big as Jupiter or even bigger. Most are also very close to their parent stars and therefore very hot.

Viewing other planets

Until recently, astronomers had no way of telling for sure if other stars had planets. Extrasolar planets are difficult to see because they are dwarfed by the light from their parent stars. However, in 1995, astronomers suggested that a star with an orbiting planet would appear to wobble very slightly when viewed through a telescope. They believed that this was because the gravitational pull of the planet would cause the star's light to bend and thus change colour. Extrasolar planets can also be detected by looking for a slight decrease in brightness as a planet comes between them and telescopes on Earth.

Space visits

With today's technology, we are unable to detect any but the closest and largest extrasolar planets. Planets the size of Earth have too small a gravitational pull to affect their parent stars enough for astronomers to notice. Even if an Earth-sized planet passed in front of its star, the drop in brightness would barely be perceptible through Earth's atmosphere. NASA's planned Terrestrial Planet Finder is due to search the brightest 1,000 stars in the Solar System between 2010 and 2020. It will be able to see Earth-sized planets and will even be able to detect what their atmospheres are like.

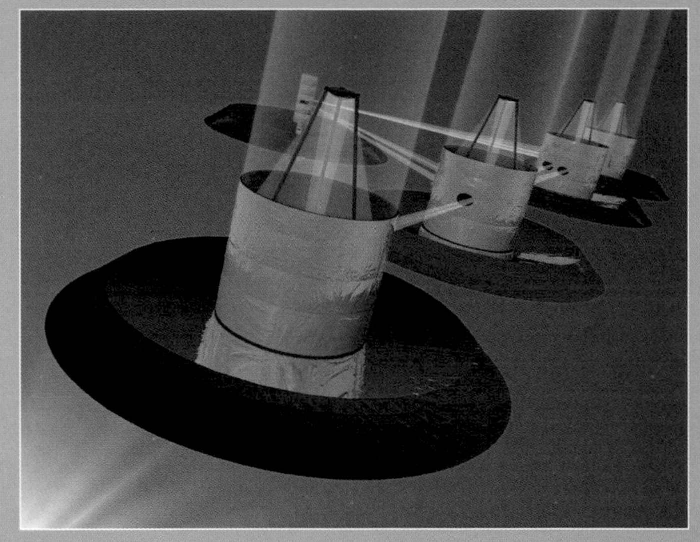

Most of the extrasolar planets found to date have highly elliptical orbits that take them very close to their suns, then very far away. This means water could not remain in liquid form.

Other Earths

In 1600, an astronomer named Giordano Bruno was burned at the stake as punishment for heresy. His crime was to suggest that around each star there might very well be planets like Earth, each with a Jesus-like figure. We now know for sure that there are potentially billions of extrasolar planets in our galaxy alone, but none discovered so far have resembled Earth. In fact, the solar systems found are shockingly different to our own, formed mainly of giant planets. Orbiting Tau Boötis is a planet four times bigger than Jupiter. It is so close to its parent star that astronomers think its clouds may be formed from vaporized rock.

Light

The Universe is unimaginably vast, and the distances between the stars so great that it takes even light, the fastest known thing, thousands of years to get from one star to the next. Because of this, the light that we see from stars is actually thousands, if not millions of years old. Many of the stars that we can view in the night sky may no longer exist, but their light can still be seen because it has taken so long to reach us.

Star light

In the early twentieth century, two astronomers named Ejnar Hertzsprung and Henry Russell created a scale for the stars. They claimed that they could be divided into two groups: main sequence stars and giants. They created a diagram to help explain this. Each star has a place on the diagram depending on how old it is. Because stars of different ages give off different coloured light, by plotting them on the HR diagram, astronomers can sort them into groups and learn more about them.

Seeing back in time

Because stars are so far apart, it takes light thousands and thousands of years to travel between even the closest of them. This means that when we look at a star at night we are gazing back in time. Even the Sun's closest neighbour, Proxima Centauri, is 4.2 light years away, which means we are seeing it as it was over four years ago. More amazingly, with powerful telescopes like the Hubble Space Telescope, we can see as far as the furthest galaxies, over 13 billion light years away. This means that we are looking back in time 13 billion years, back to the beginning of the universe!

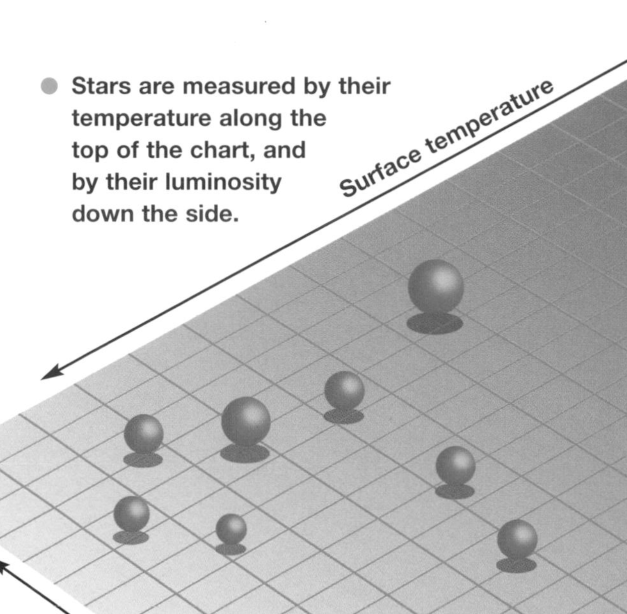

Stars are measured by their temperature along the top of the chart, and by their luminosity down the side.

Most stars lie in a band stretching from top left to bottom right. This is the main sequence. White dwarfs lie along the bottom of the diagram because they are still very hot but are not as bright.

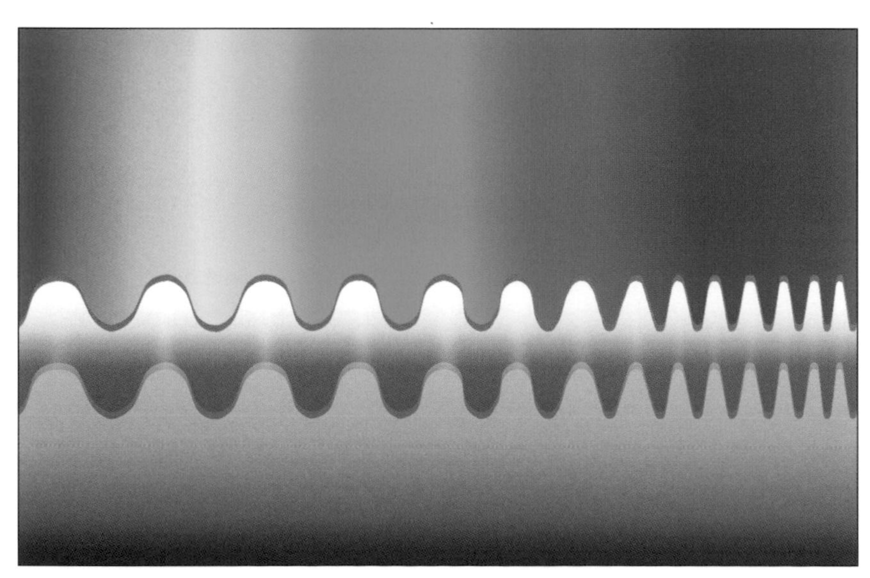

Light travels through space like a ripple in a pond or a wave on the ocean. The distance from one crest to the next is known as the wavelength of the light. Visible light ranges from blue (which has short wavelengths) to red (which has long wavelengths).

Star colours

If you look closely at the stars in the night sky, you will see that they glow in a variety of different colours. This is because they have different temperatures and emit light with different wavelengths. Very hot stars, with temperatures greater than 28,000°C (50,400°F), glow blue, and slightly cooler stars, with temperatures of around 7,000°C (12,600°F), appear to be white. Stars like our Sun, which have a surface temperature of about 5,500°C (9,900°F), are yellow. Cooler stars with temperatures lower than 3,000°C (5,400°F) seem red in the night sky.

Red giants and supergiants lie along the top of the chart because, although they are much cooler than newer stars, they are much larger and much brighter.

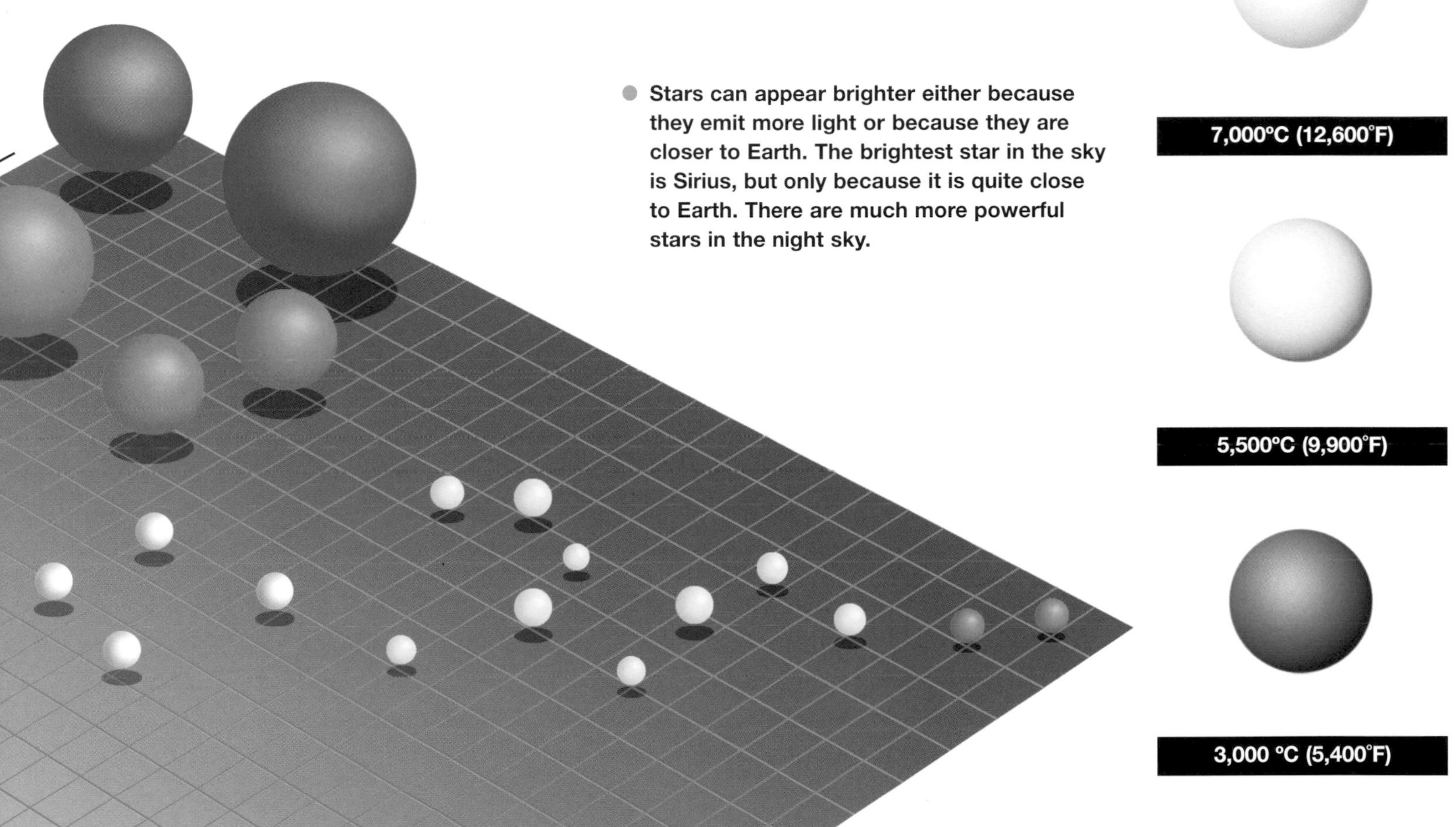

Stars can appear brighter either because they emit more light or because they are closer to Earth. The brightest star in the sky is Sirius, but only because it is quite close to Earth. There are much more powerful stars in the night sky.

28,000 °C (50,400°F)

7,000°C (12,600°F)

5,500°C (9,900°F)

3,000 °C (5,400°F)

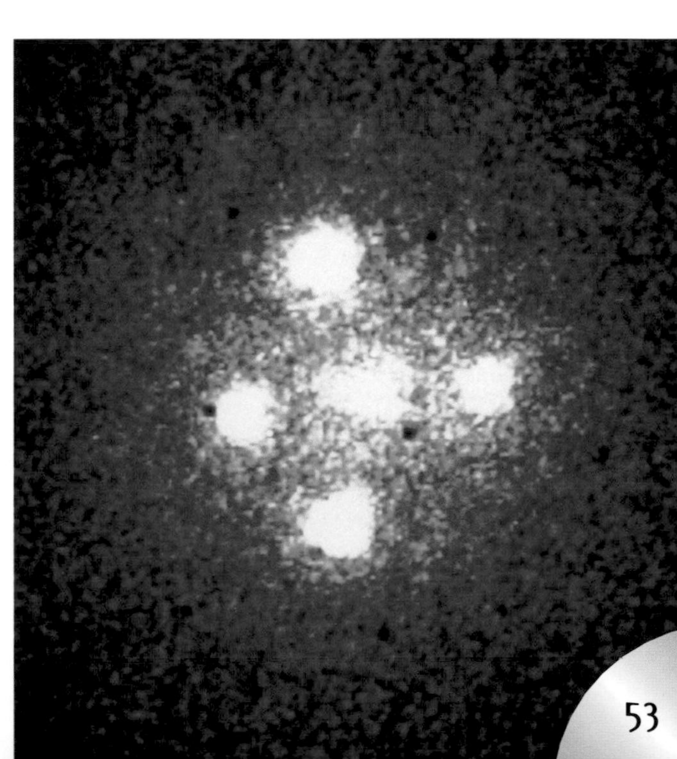

Light can be bent by gravity. The Einstein Cross is actually only one quasar, very far away in the Universe. However, its light is bent on its way to Earth by a galaxy so that it appears as five points of light.

Black Holes

Black holes are the monsters of the Universe. Formed from the brightest of all explosions, supernovae, they soon become the darkest objects in space, emitting no light at all. The gravitational force of a black hole is so powerful that nothing, once caught, can escape. Even light can be pulled off course and into a black hole, where it is trapped for ever.

Neutron stars

When a star explodes into a supernova, it does not always die away but sometimes gives birth to something new. The incredibly strong gravitational pull of the star means that the core collapses in on itself to form a very small, very dense ball. If the star contains less than three times as much matter as the Sun, it becomes a neutron star. Its matter is packed incredibly tightly into a space only a few kilometres in diameter. It is so dense that a single thimbleful can weigh more than the Eiffel Tower!

- **If you were to drop a small amount of the matter from a neutron star onto the surface of the Earth, its weight would be so great that it would sink through the crust, down to the planet's core!**

Black holes

If the star's core after a supernova is more than three times the mass of the Sun, it will collapse even further than a neutron star, shrinking itself into an unimaginably small point called a singularity. Its gravity becomes immensely strong, creating a "gravitational well" in space. Anything that passes too close to this well is sucked into it for ever. The force is so strong that nothing can escape, not even light.

Black holes emit no light. However, scientists can find them if they are located close to another star. The pull of a black hole will tear gas from a nearby star, as shown below. This gas will circle the black hole with such force that its temperature can exceed 100 million °C! This is so hot that X-rays will be released as in the image on the left, allowing astronomers to pinpoint the black hole's location.

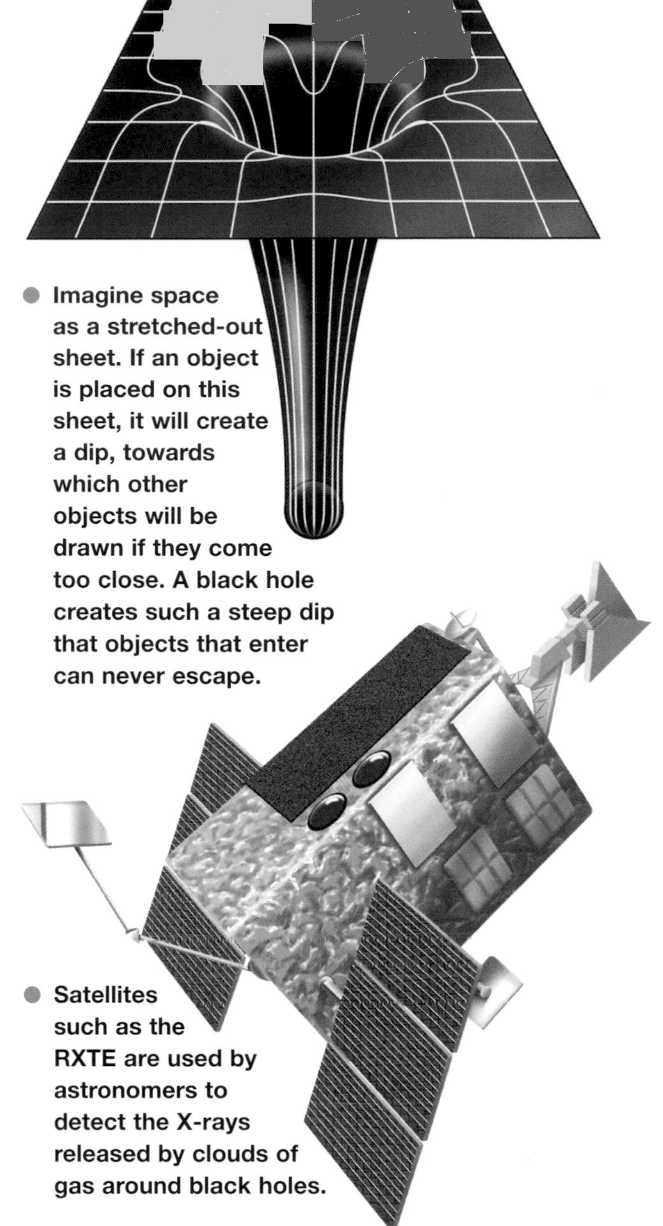

Imagine space as a stretched-out sheet. If an object is placed on this sheet, it will create a dip, towards which other objects will be drawn if they come too close. A black hole creates such a steep dip that objects that enter can never escape.

Satellites such as the RXTE are used by astronomers to detect the X-rays released by clouds of gas around black holes.

As a star contracts, it begins to spin faster, turning many times a second and sending beams of radiation across space like a lighthouse. When a star does this, it is called a pulsar.

Gateways

Some scientists believe that black holes are portals that may connect some parts of the Universe with others. Nobody really knows how this would work, but science fiction has long suggested the idea of a worm hole, an artificial black hole that would act as a tunnel through space and time. Black holes are believed to "end" at a singularity inside, where everything is compressed. However, if a tunnel could be held open by an anti-gravitational force, then light and matter would not be trapped but able to pass freely from one part of space to another.

Galaxies

Galaxies are immense collections of stars held together by gravity. While individual stars may seem vast in comparison with Earth, galaxies can often contain trillions of stars and span hundreds of thousands of light years. Scientists are still unsure how galaxies formed from the dark void of fog and gas that was the early Universe.

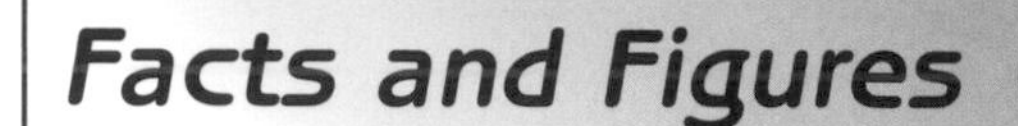

Facts and Figures

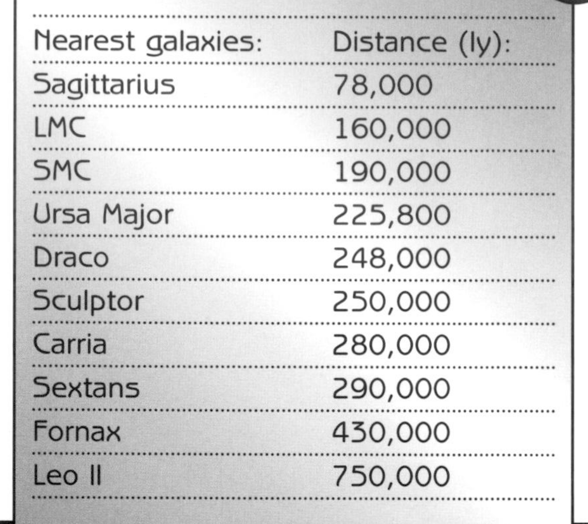

Nearest galaxies:	Distance (ly):
Sagittarius	78,000
LMC	160,000
SMC	190,000
Ursa Major	225,800
Draco	248,000
Sculptor	250,000
Carria	280,000
Sextans	290,000
Fornax	430,000
Leo II	750,000

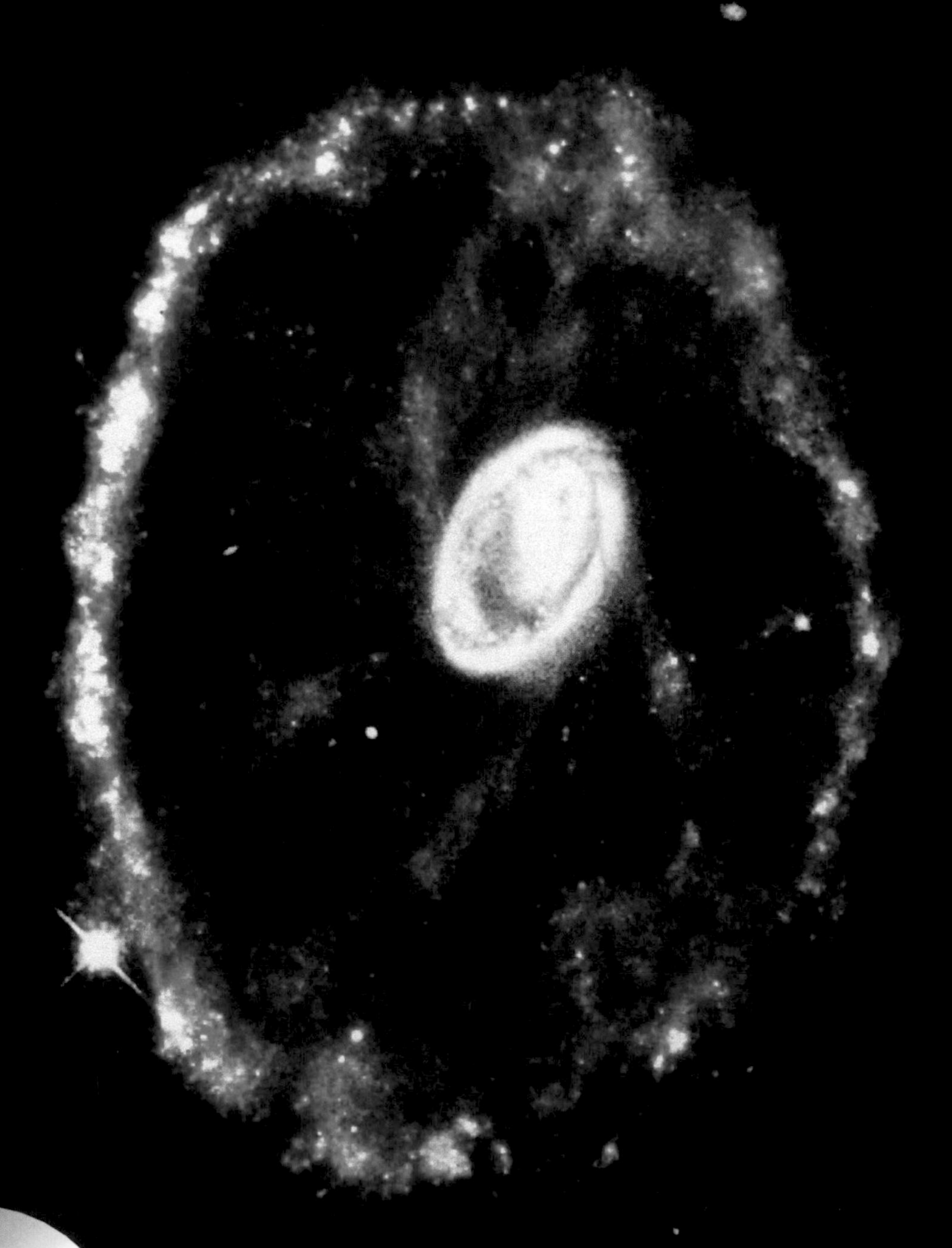

- **Galaxies are also known as island universes after a suggestion by Immanuel Kant, a German philosopher.**

- **Most galaxies form when giant clouds of gas collide. If the clouds are rotating, they form a spiral galaxy. If not, they usually form an elliptical galaxy.**

- **The Cartwheel Galaxy was hit by another galaxy around 300 million years ago. It used to be a spiral galaxy, but a smaller galaxy, travelling very fast, smashed through its centre. The blue ring around the central region is made up of millions of new stars triggered by the impact.**

Galaxy shapes

Galaxies come in all shapes and sizes. A third of all galaxies are spiral, spinning around space like Catherine wheels. Old stars are packed tightly in their centre, while young stars emerge continuously from the gas and dust in their spiral arms. Elliptical galaxies get their name from their oval shape. They are formed from billions of old red stars and have little gas with which to create new stars. Some galaxies cannot be classed as either spiral or elliptical as they have no recognizable shape. They are called irregular galaxies and are usually quite small. These galaxies are full of gas in which new stars are forming all the time.

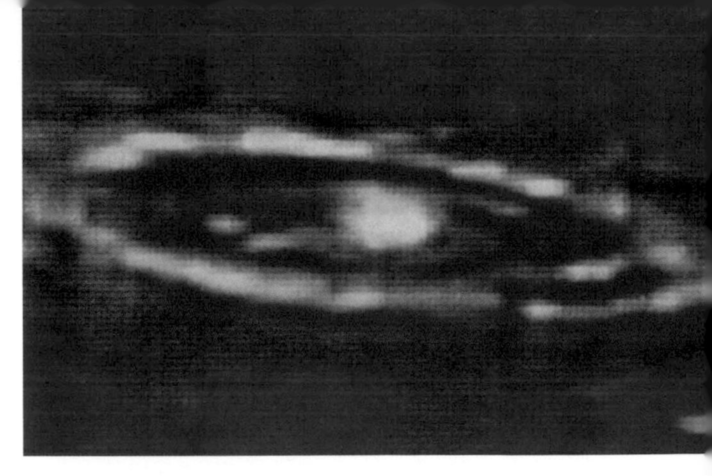

A spiral galaxy (viewed in infrared)

An elliptical galaxy

Cosmic collisions

Galaxies are normally separated by vast, empty gulfs. However, very rarely two galaxies pass close enough to each other to collide. The galaxies are travelling at millions of kilometres per hour, and the impact can be tremendous. The individual stars in a galaxy do not collide, but the vast clouds of interstellar gas and dust smash into one another, triggering a ferocious birth of new stars. The results of an impact vary. Each galaxy is held together by nothing but gravity, causing an immense battle of strength. Sometimes galaxies merge to form even larger galaxies. At other times, galaxies can be distorted or even ripped apart by the impact.

An irregular galaxy

Active galaxies

A small percentage of the galaxies known today stand out from the rest. Violent explosions occurring in their cores force out giant streams of gas. These plumes can stretch for thousands of light years. Such galaxies are called active galaxies (shown left), and their energy does not come from the light of the stars inside them but from the heat generated by enormous black holes in their central regions. Quasars are among the most powerful and most distant galaxies in the Universe. Their light has taken billions of years to reach us, so they show us how the Universe behaved near its birth.

The Milky Way

If you look up at the sky at night, you can often see a band of stars that stretches across the sky like a river of milk. By viewing closely through a pair of binoculars, you will see that it is actually a dense collection of stars. This is the Milky Way, and what you are seeing is our own galaxy from the inside. Our Sun is one of over 200 billion stars that make up our galaxy.

Facts and Figures

Diameter: 100,000 light years
Depth: 2,000 light years
Mass: 1,000 million million solar masses
Time taken to rotate: 220 million years

The Milky Way spins more slowly in its densely packed centre and its spiral arms than it does in its outer regions. Its average speed is around 230 kilometres per second (143 miles per second).

The Local Group

Our galaxy may seem to be immense and of vast importance, but even the Milky Way is not isolated in space. It is only one of approximately 30 galaxies that make up the *Local Group*, an enormous collection of galaxies that stretches over millions of light years. Most of the galaxies in the Local Group are clustered around the strong gravitational pulls of the Milky Way and the similar Andromeda Galaxy. However, just as galaxies group together to form clusters, clusters are attracted to each other to form superclusters! These are immense clouds of galaxies that can stretch for hundreds of millions of light years.

Our Solar System lies halfway along the Orion arm of the Milky Way. The Sun takes 220 million years to orbit the centre of the galaxy.

The centre of the Milky Way is marked by a tiny radio source called Sagittarius A*. This is emitted by the gas around a black hole.

The Milky Way

Known as the Milky Way, our galaxy contains over 200 billion stars. It looks like a giant spiral from above (left), but if it was viewed from the side it would appear as a flat band of stars (right, seen in infrared). This is because although the Milky Way is 100,000 light years long, it is only 2,000 light years thick. Our Solar System lies halfway along one of the spiral arms. Scientists believe that in the centre of the Milky Way there is an enormous black hole with a mass greater than 2.5 million Suns. It is not active at the moment, but it may come back to life again!

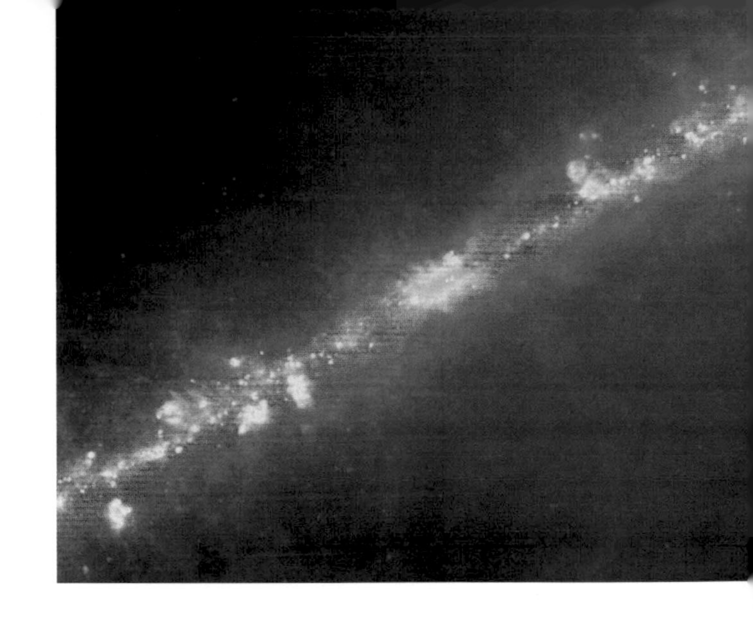

The Milky Way can be seen most clearly in both hemispheres between June and September. It appears as a band because the galaxy is flat, and we are viewing it from inside.

Galactic partners

Orbiting our galaxy, just as the Earth orbits the Sun, are two enormous satellite galaxies called the Magellanic Clouds. The Large and the Small Magellanic Clouds consist of thousands of stars and orbit the Milky Way every 1.5 billion years. The Large Magellanic Cloud (below) is made from the same mix of gas and stars as the Milky Way, but it is less than one twentieth as large and too small to grow spiral arms. It lies 160,000 light years away. The Small Magellanic Cloud is 190,000 light years away and is slowly being pulled apart by the gravity of the Milky Way.

Between the Stars

The space between the stars may appear black but it is certainly not empty. A handful of space contains, on average, 30 hydrogen atoms and several specks of dust. This may not seem much, but it is this interstellar medium that is constantly giving birth to new stars. Light from these new stars shines through the clouds to produce incredible shapes called nebulae.

Interstellar medium

The space between the stars in a galaxy is not empty. It contains small quantities of dust and gas called interstellar medium. This gas is what remains of stars that have exploded. It will in turn give birth to new stars when the conditions are right. Most interstellar medium can be found in clouds and can rise to temperatures greater than 8,000°C (14,400°F). However, there are pockets of very thin gas between the stars where temperatures can reach higher than one million °C (1.8 million °F)! These are bubbles created by the force of supernovae.

Nebulae

Interstellar medium has an average density of less than one atom for every cubic centimetre, but in some places it is much more closely spaced than this. A great deal of interstellar dust and gas is concentrated into vast clouds called nebulae. These nebulae are both cosmic nurseries and graveyards in space. New stars form from the gas and dust in a nebula, and it is the light of these stars that gives the clouds their colour. However, when many stars die, they blow off their outer layers to form a planetary nebula (see opposite page), sending material back into space, where it will eventually form new stars.

- **Even in nebulae the interstellar medium can only amount to several hundred atoms per cubic centimetre. We can still see interstellar medium because we are looking through a great thickness of it.**

- **Much of the starlight in the Milky Way is blocked out by clouds of gas and dust. If these clouds were not there, the light from the stars at night would be bright enough to read a book by!**

Types of nebulae

Nebulae come in many different shapes, sizes and colours. Most are illuminated by the light of the stars inside them, but some do not give off any light at all and can only be seen as dark shadows that block out the light of distant stars.

Emission nebulae

These are the most beautiful and colourful of all the nebulae. Their striking colours come from the presence of hydrogen atoms that release red light. These clouds are very hot, often over 10,000°C (18,000°F).

Reflection nebulae

Reflection nebulae are illuminated by light that is reflected from nearby stars. They appear blue because the light is scattered by dust grains. This produces the same effect as the Sun's light shining through the Earth's atmosphere and making the sky appear blue.

Dark nebulae

These nebulae, also called absorption nebulae, appear dark because there are no nearby stars to light them. They can be spotted because they blot out the light from more distant stars, appearing like black voids in the sky.

Planetary nebulae

After stars like our Sun become red giants and die, they blow off their outer layers in a large, expanding cloud. Early astronomers believed these clouds looked like Uranus and Neptune, but now we know they have nothing to do with planets.

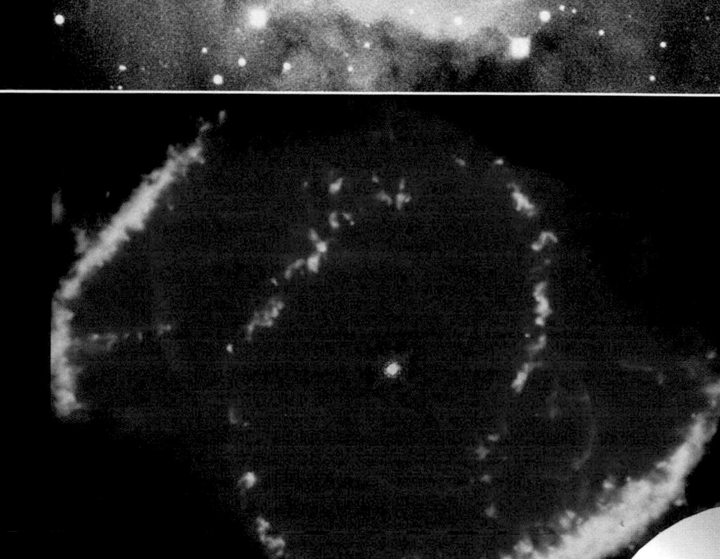

In the Beginning

Nobody knows how the Universe began, but the most common theory is the Big Bang. According to this, the Universe was formed from an immense explosion 13 billion years ago. Before the Big Bang, everything in the Universe was packed into a tiny area, smaller than the nucleus of an atom. This point was called a singularity and was incredibly hot. It was released in an explosion so powerful that all of the matter in the singularity was blasted into an area larger than a galaxy in a fraction of a second.

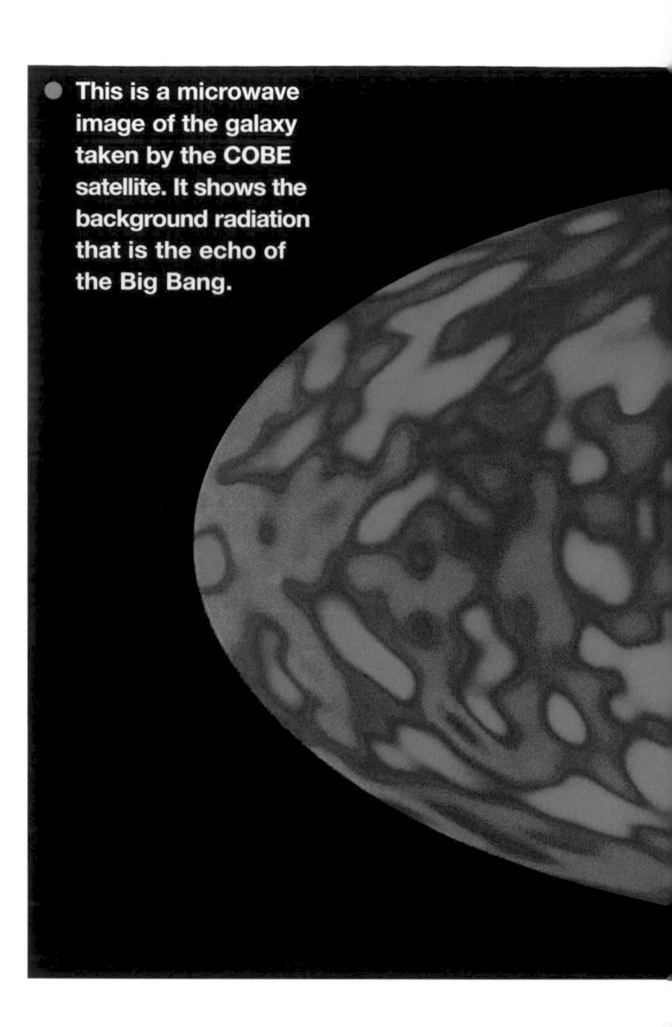

This is a microwave image of the galaxy taken by the COBE satellite. It shows the background radiation that is the echo of the Big Bang.

How it happened

The Big Bang was an unimaginably powerful explosion of energy that sent particles flying in every direction. A process called inflation meant that it grew from the size of a minuscule particle to an area bigger than the whole Milky Way in a fraction of a second. Moments later, it began to cool into a soup of atomic particles.

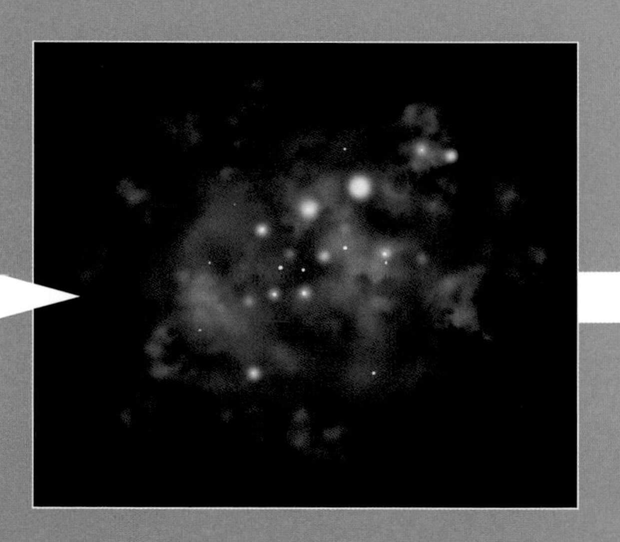

The atomic fireball cooled further as it expanded, changing into a foggy cloud of matter. This cloud was so dense that even rays of light could not travel far inside it. Energetic, unstable atomic particles began to clump together within the first three minutes, forming more stable groups. These were the building blocks of everything in the Universe.

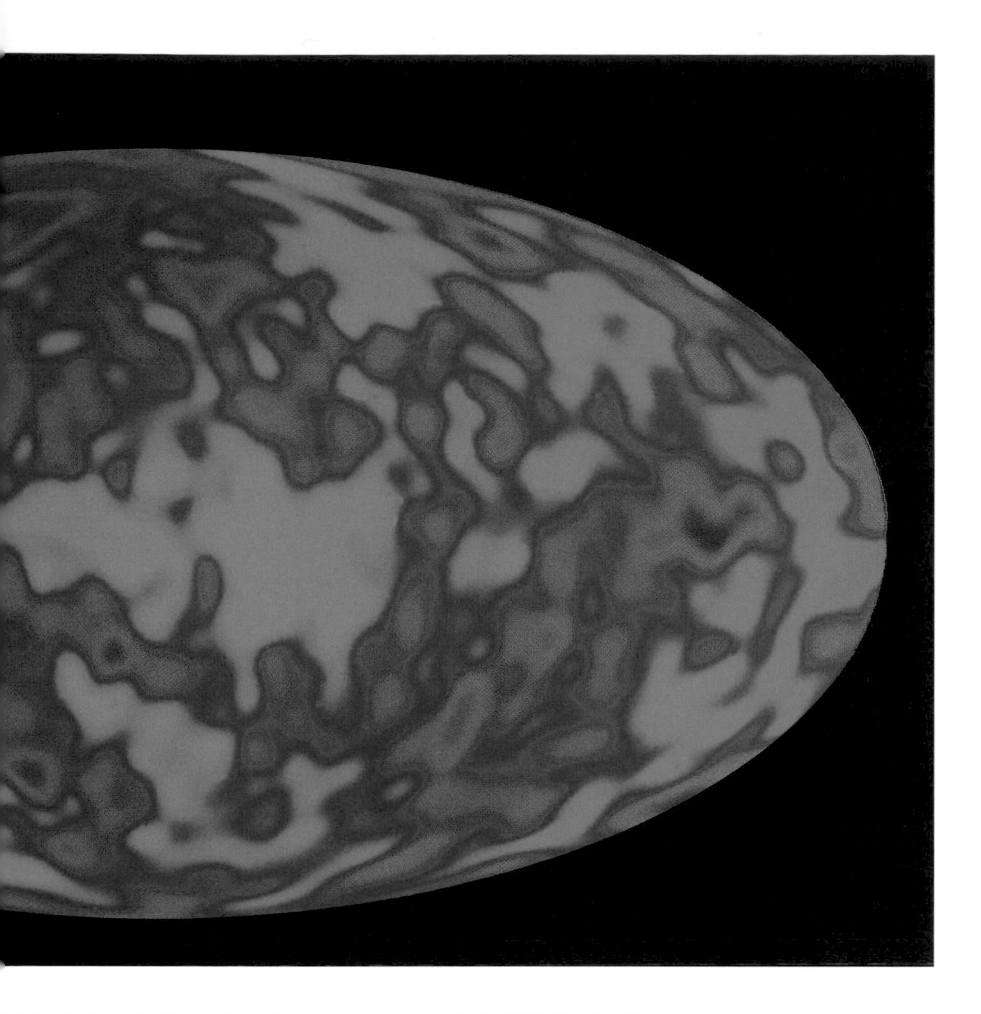

After thousands of years, the temperature in the Universe fell to a few thousand degrees Celsius. This change in temperature caused the atomic particles to alter their form, and the thick fog cleared away. Galaxies of stars began to form, and the Universe became as transparent as it is today.

Evidence

There is very strong evidence to support the theory of the Big Bang. The strongest proof is a weak signal that has been detected in space. This is thought to be an echo from the energy released by the force of the Big Bang. In 1992, the Cosmic Background Explorer, or COBE satellite (below), found weak ripples in the background radiation of space. Scientists think this may be the afterglow of the Big Bang. However, for the Big Bang theory to be correct, the Universe must contain more matter than we know about. If the Universe contains merely the matter we know, its expansion would have happened too quickly for galaxies to form.

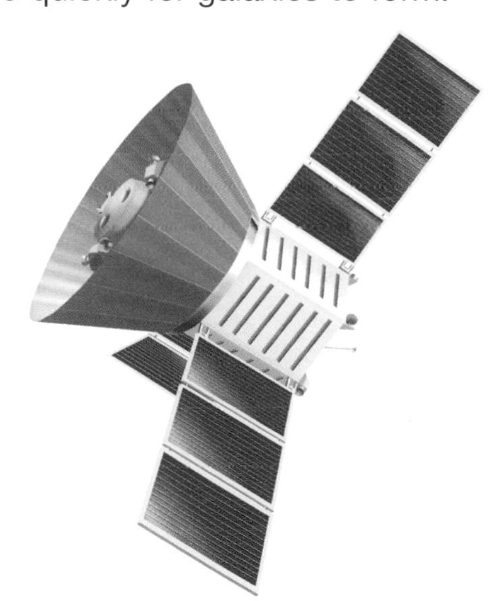

Other theories

There have only been a handful of scientific alternatives to the Big Bang theory. The most famous of these was the Steady State theory, which suggested that the Universe has no beginning and no end, and that though the Universe is expanding, it stays in perfect balance over time, like a washing-up bowl kept topped up by a tap. Nowadays, the Steady State theory has been dismissed by most scientists because of the abundance of proof for the Big Bang.

Creation Myths

The Big Bang theory is a scientific explanation of how the Universe began. Additionally, cultures across the world, from before the ancient Babylonians to modern times, have had their own theories about how the Universe, and life within it, began. These myths have helped people understand the big questions of life for thousands of years.

In the beginning

Some of the greatest questions to trouble humankind for thousands of years concern the beginning of the world and the origins of life. Scientists today are learning more and more about the Earth, the Solar System and the Universe beyond, and are putting forward their own theories about how it all began. However, since the earliest humans began to study the world around them, they have formed mythologies to help answer these big questions.

America

According to the Native American Madoc tribe, the Old Man of the Ancients, whose name was Kumush, created the world. Kumush journeyed to the end of the Earth and returned with a daughter. He made her ten dresses, the tenth of which was her burial dress and the most beautiful. When she died, Kumush followed her to the House of Death, where they stayed for many years. Eventually, he returned to the sunlight with a basketful of bones, which he planted. These became the first tribes, and Kumush became the Sun.

Egypt

According to ancient Egyptian mythology, the fundamentals of life—air (Shu) and moisture (Tefnut)—came from the spittle of Re, the Sun god. From the union of Shu and Tefnut came Geb, the Earth god, and Nut, the sky goddess. The first human beings were born from Re's tears.

China

In the beginning of time, according to Chinese mythology, everything was chaos, and the chaos was in the form of a chicken's egg. The two opposing forces of Yin and Yang lived inside this egg, constantly fighting, until one day their violence split the egg in two. Heavy elements inside the egg became the Earth, and lighter elements became the sky. In between was the first being, P'an-ku, shown on the right. Humans were born from P'an-ku's fleas!

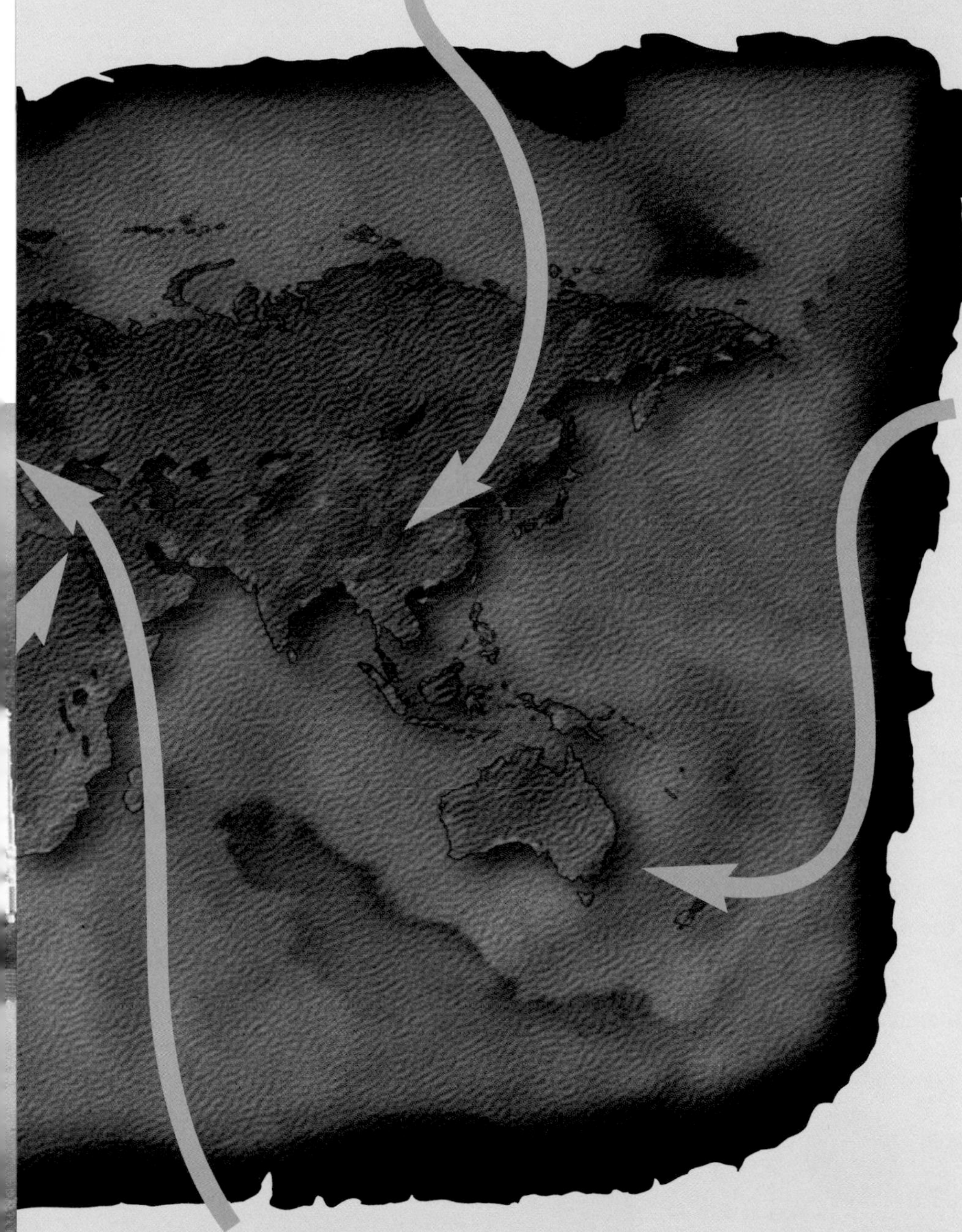

Australia

According to the mythology of Australian Aboriginals, the Earth started life as a bare plain, with the Sun, the Moon, the stars and the Eternal Ancestors sleeping beneath the Earth. The Eternal Ancestors rose from their slumber and roamed the Earth in different forms, sometimes as animals and sometimes as plants. Two self-created beings, the Ungabikula, found many unfinished people made from pieces of plant and rock. Using their knives, they finished carving the people, and humankind was born.

Serbia

Serbian mythology has it that the Universe was born when God, who had always slept, awoke. As he looked around himself, stars and planets formed wherever his eyes fell. God explored the celestial bodies that he had created until he was exhausted and drenched in sweat. When he arrived on Earth, a drop of sweat fell onto the soil and created the first human.

Visitors from Space

While scientists are busy trying to uncover signs of life from the furthest reaches of our galaxy, thousands of people here on Earth are convinced that we have already been visited by extraterrestrial beings. Hundreds of people every year report seeing unidentified flying objects (UFOs) such as flying saucers. Some even claim to have been abducted by aliens and carried into space!

UFOs

Hundreds of thousands of reports of mysterious objects in the skies are made every year. Over 95% of these can be explained in terms of everyday things such as weather balloons and strange cloud formations. However, 5% cannot be explained, and it is these sightings that many believe to be visiting aliens. In 1948, a craft was seen by several witnesses crashing next to a military base in Roswell, USA. A strange metallic material was found next to the crash site. The metal was very thin yet could not even be bent with a sledgehammer and was resistant to extreme heat.

- **The term "flying saucer" was used by a farmer in Texas as far back as 1878 to describe what he thought was a UFO. The US Army built its own flying saucer in the 1950s, but it barely flew!**

Signs of life

In December 1990, the Galileo probe began to investigate a planet that displayed a multitude of signs of life. This planet was Earth, and the probe was conducting a very interesting experiment: whether it is possible to detect signs of life on a planet when viewed from space. The Galileo probe detected that Earth had both water on its surface and oxygen in its atmosphere, which told scientists that this planet contained life. Any visiting life forms would also be able to detect that Earth is home to living, breathing organisms. As well as this, the Earth at night is a lighted signpost pointing alien spacecraft in our direction!

Earth at night is like a glowing, neon signpost. Lights are brightest in the big cities but can also come from burning oil wells and large fleets of boats.

Oxygen is a very reactive gas and would quickly disappear from the atmosphere if it was not constantly being replenished by living plants. Aliens detecting oxygen in the atmosphere would know there was life on Earth.

Alien life forms

Many people have claimed that alien life forms have visited Earth. Some think that superintelligent aliens landed on Earth 40,000 years ago and created human beings. Many even claim that these extraterrestrial beings returned several times to teach humans the secrets of technology so that they could build wonders such as the Great Pyramids. Nobody knows what would happen if aliens did visit Earth today. Would they be friendly or hostile? In 1938, a radio adaptation of H G Wells' novel *The War of the Worlds* led many terrified listeners to flee New Jersey, USA, because they thought it was being invaded by hostile Martians!

The End

Our Solar System—indeed, the whole Universe—will not live for ever. Scientists now know that in billions of years' time the inner planets, including the Earth, will be destroyed by the dying Sun. The Universe itself also has a limited life span. It will either continue to expand for ever, dying slowly until it is nothing but a thin mist, or collapse back into a single particle, possibly to be reawakened again in another Big Bang.

The Sun's future

Like all stars, our own Sun was born and will eventually die. Scientists estimate that it is currently halfway through its immense life cycle, quite stable in its main sequence. However, in five billion years the Sun's supply of hydrogen will run dry, and it will transform into a red giant as it burns helium instead. It will eventually expand to over thirty times its present size, engulfing the inner planets, including Earth and Mars. When the helium runs out, the Sun will cast off its outer layers to form a planetary nebula, after which its core will remain as nothing more than a tiny white dwarf.

Conditions on Earth will become unbearable as the Sun expands. Humankind will have no choice but to leave in search of a new home.

A new home

Humans have so far only visited one alien world in person—the Moon. However, as the Sun begins to die and expand, conditions on Earth will become too hot for humans to survive, and our descendants will be forced to leave this planet in search of a new home. Mars might very well be the first port of call, and could provide a home for humankind for a few million years. However, Mars too will eventually be engulfed by the dying Sun. Humans will be forced to travel through space to find an Earth-like planet orbiting a younger star.

The Sun will swell up as it burns the last of its fuel. At this point, it will already have engulfed Mercury and Venus. Earth and Mars will be next.

Dark matter

Scientists cannot see all of the matter that makes up space, even with the most sophisticated telescopes. There is a great deal of matter in the Universe that cannot easily be spotted from Earth. This is called dark matter. It includes objects such as brown dwarfs and black holes. Scientists know that this dark matter exists because its gravitational pull can be seen on visible objects such as stars and galaxies. Dark matter is thought to make up an incredible 95% of the mass of the Universe!

The end of everything?

The key to the future of the Universe lies in gravity. If the Universe contains enough mass in the form of visible and dark matter, then eventually gravitational forces will cause its current expansion to slow, then stop for a split second before beginning to contract. It will collapse in on itself, shrinking faster and faster until there is a "big crunch". This is the opposite of the Big Bang, when all of the matter in the Universe collides in the centre of an immense black hole. Alternatively, if there is not enough matter in the Universe to stop it from expanding, it will do so for ever, its stars gradually decaying until nothing remains but a mist of particles.

A great deal has changed since humankind first looked up at the stars. What our early ancestors were only able to gaze up at with awe and wonder, some people can now visit in person. Space probes have visited all eight planets, and one has even looked inside a comet. Hundreds of people have flown in space, and many have even lived inside a space station floating hundreds of kilometres above the surface of the Earth. These incredible developments in space exploration have come about because of the enthusiasm and hard work of countless scientists and astronomers. Their work today may eventually enable humankind to travel beyond the Solar System to the stars themselves.

Exploring the Solar System

Early Astronomy

Ever since humankind first looked at the sky, it has been fascinated by the heavens. Many ancient cultures saw the stars as the abode of the gods and attributed personalities to celestial objects and movements. The curiosity of the ancient Greeks taught us a great deal about the depths of space, but we are still a long way from knowing the secrets of the Universe.

Stonehenge is a giant astronomical calendar with stones aligned to the Sun. It was built around 3000BC.

First skywatchers

Evidence has recently been found to suggest that humankind has been charting the skies for over 15,000 years. Cave paintings found recently in France and Spain include maps of star clusters such as the Pleiades. The Akkadians kept astronomical records over 4,500 years ago. There is evidence to show that they predicted the course of objects in the sky, including the Sun, the Moon and the planets. Fascinating ancient sites such as the Pyramids in Egypt and Stonehenge in England are thought to have astronomical significance.

Ptolemy

It was the ancient Greeks who turned astronomy into a science. Ptolemy published his *Almagest* in AD140. This was a remarkable encyclopedia of the patterns of the stars and the planets. He used it to support his argument that the Earth was the centre of the Universe. His "System of the World" suggested that surrounding Earth were seven transparent spheres, each containing a moving object. He claimed that an eighth sphere surrounded everything, and that the stars were points of light set on this sphere. Although we now know this theory to be wrong, it was a very accurate method for predicting the motions of the planets, and was the cornerstone of astronomy for 1,000 years.

Facts and Figures

- 13,000BC: Cave-paintings in Europe show evidence of skywatching.
- 750BC: The Babylonians work out the cycle of the Moon.
- 164BC: The Babylonians make the first recorded sighting of Halley's Comet.
- AD150: Ptolemy claims that the Earth is the centre of the Universe.
- AD1054: The Chinese record the Crab supernova exploding.
- AD1543: Copernicus suggests that the Earth revolves around the Sun.

- **Ptolemy believed that the Earth was the centre of the Universe. Moving outwards from Earth was the Moon, Mercury, Venus, the Sun, Mars, Jupiter and Saturn. The eighth sphere held the stars.**

- **Ptolemy also believed that the sky could never change its shape. An astronomer named Tycho Brahe proved this wrong in 1572 when he saw a supernova and a comet.**

Copernicus

Aristarchus, another Greek astronomer who lived in Ptolemy's time, was one of the first to suggest that the Sun was at the centre of the Universe, and that the Earth orbited it. His ideas were laughed at, and it was not until fifteen hundred years later that Ptolemy's theory was challenged seriously. In the 1540s, a Polish churchman named Copernicus claimed that the planets, including Earth, orbited the Sun. He was arrested by the Church for his ideas, but eventually the work of astronomers such as Galileo proved him right.

- **Copernicus believed that the Sun was the centre of everything in the Universe. His system placed the planets around the Sun and is very similar to the Solar System we know today.**

All in the Stars

From the earliest times, people have seen images in the patterns formed by the stars. Societies all over the world have grouped the stars into constellations and projected characters, animals and other beings into these groupings. These characters have become so important to cultures that many myths and stories surround them. Some people today still believe that the position of the stars can influence their lives.

The stars appear to move in the night sky because Earth is in constant motion around the Sun. The image above shows how the stars appear to move over a long time.

The ecliptic is the line across the sky through which the Sun appears to move. Nowadays, we know that it is actually the Earth that moves around the Sun. However, many years ago people believed that the Sun was travelling through these twelve constellations.

The Zodiac

Of the 88 constellations that are recognized today, the most ancient form the Zodiac. This is a band of twelve constellations that lie along the ecliptic, the part of the sky through which the Sun appears to move during Earth's orbit. Because the Sun always travels through the same part of the sky, these constellations were seen as special in ancient times. Even today, they act as markers in astrology, the belief that human life is influenced by the position of the stars and planets.

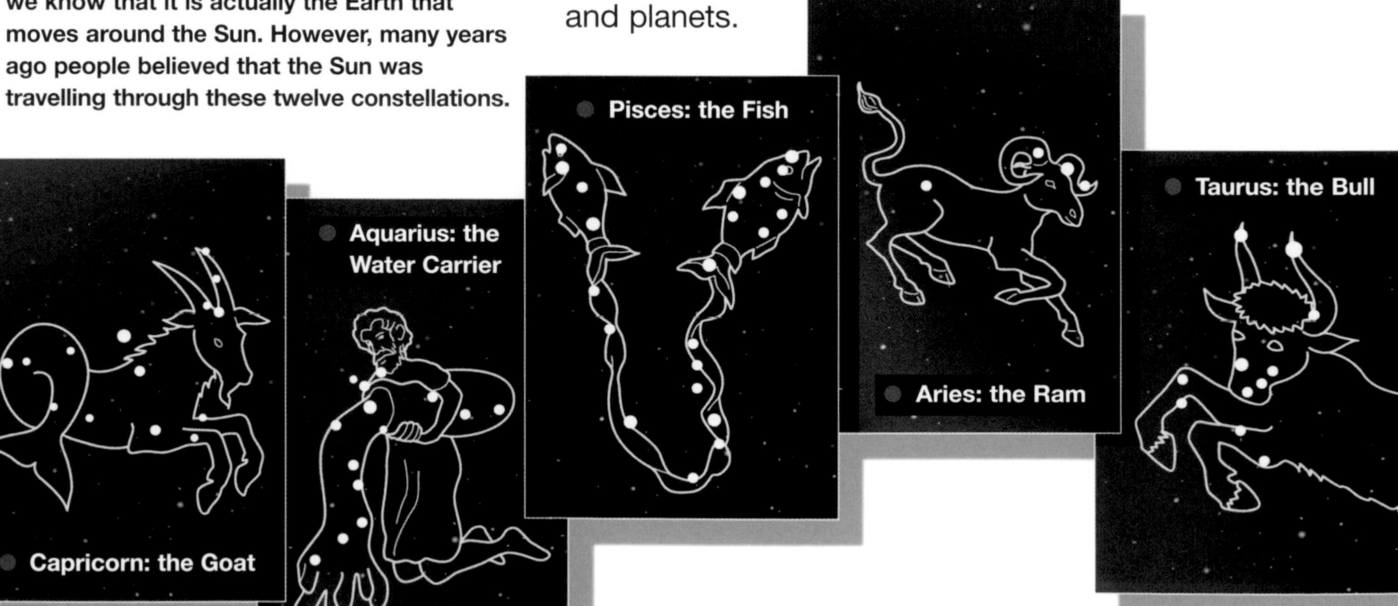

Capricorn: the Goat

Aquarius: the Water Carrier

Pisces: the Fish

Aries: the Ram

Taurus: the Bull

Facts and Figures

- Aries: March 21 – April 20
- Taurus: April 21 – May 21
- Gemini: May 22 – June 21
- Cancer: June 22 – July 23
- Leo: July 24 – August 23
- Virgo: August 24 – September 23
- Libra: September 24 – October 23
- Scorpio: October 24 – November 22
- Sagittarius: November 23 – December 21
- Capricorn: December 22 – January 20
- Aquarius: January 21 – February 19
- Pisces: February 20 – March 20

Constellations

There are thousands of stars visible in the night sky, so many that on a clear night you would be able to pick out almost any shape you liked! The most prominent stars in the sky have been grouped together by skywatchers over thousands of years to form patterns called constellations. These range from animals, such as the Great Bear, to mythological heroes, such as Perseus. There is usually no real link between the individual stars of a constellation. Although astrologers draw lines between the points of light to identify shapes in the night sky, the individual stars in any constellation can be hundreds of light years apart.

Mapping the sky

Knowing the patterns in the night sky is as useful to an astronomer as knowing the layout of a new town is to a tourist! It is much easier to get to grips with the innumerable mass of stars in the heavens if you have markings to go by. Astronomers are not the only people to use the constellations as a map. Ever since humankind first ventured away from the land in boats, they have been safely guided by the stars. Sailors around the world would learn the position and movements of the constellations to help them navigate their way across the oceans.

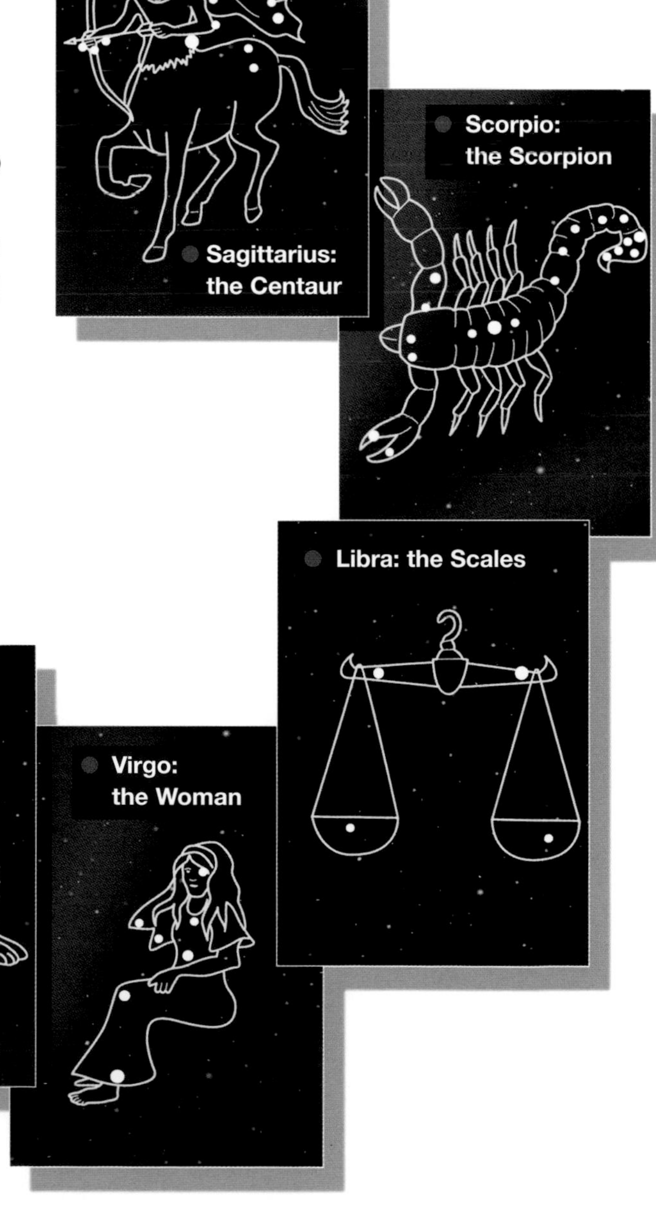

Cancer: the Crab

Gemini: the Twins

Telescopes

Telescopes were by far the most important invention in astronomy. With the naked eye, early astronomers could see only a few thousand of the stars in the night sky. However, when Galileo first turned his homemade telescope to the skies, he found that he could see craters on the Moon, the stars in the Milky Way and even four moons orbiting Jupiter!

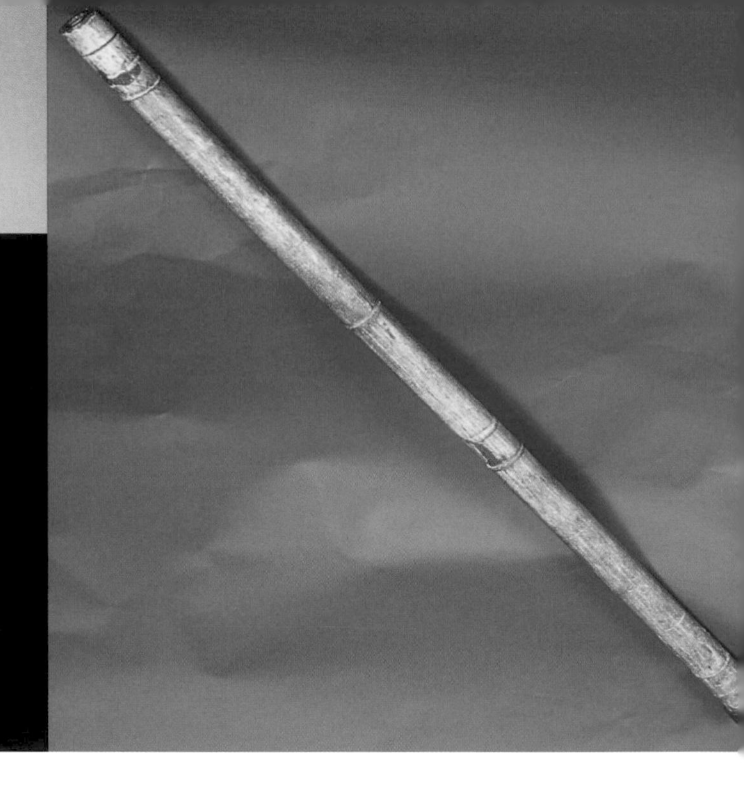

Early telescopes

The most important revolution in astronomy came in 1608 when the Dutch spectacle-maker Hans Lippershey invented the first telescope. It was a simple design that only magnified objects a small number of times. However, the idea spread like wildfire across Europe. Galileo Galilei used the design to produce the first astronomical telescope a year later (shown above). His telescope was not much more powerful than Lippershey's and no more powerful than a pair of modern binoculars. However, through it Galileo discovered many amazing things about the Solar System, including Saturn's rings and four of Jupiter's moons.

- **When Galileo (shown left) discovered four moons orbiting Jupiter, he began a process that revolutionized the way people thought about the Solar System. Everything did not, after all, orbit the Earth. Could it really be at the centre of the Universe?**

- **The main image in a reflecting telescope is formed in the main mirror. A secondary mirror is needed to direct this image to the viewer.**

- **The secondary mirror reflects the light of the main image to an eyepiece, or to data-recording equipment.**

Facts and Figures

1608: The first refracting telescope is invented by Dutch spectacle-maker Hans Lippershey.

1610: Galileo Galilei uses a refraction telescope to view sunspots, craters on the Moon and moons around Jupiter.

1655: Christiaan Huygens uses a telescope to view Saturn and suggests that it is surrounded by a ring. He also discovers Saturn's moon, Titan.

1663: The first reflecting telescope is invented by Scottish astronomer James Gregory.

1845: The first large telescope is built by Irish astronomer the Earl of Ross. It has a 180-cm (6-ft) mirror.

How they work

There are two different types of astronomical telescope: reflectors and refractors. Both work by capturing as much light as possible from distant objects such as planets and stars and directing that light to the human eye, or to data-recording equipment such as computers. Refractor telescopes use lenses to capture light, which is then focussed into an image by a second lens. Binoculars usually consist of two refracting telescopes side by side. Most professional astronomers prefer to use reflector telescopes, which have curved mirrors to capture light. The main lenses of all telescopes have a much larger surface area than the human eye, which means that more light can be collected, and objects can be seen more clearly.

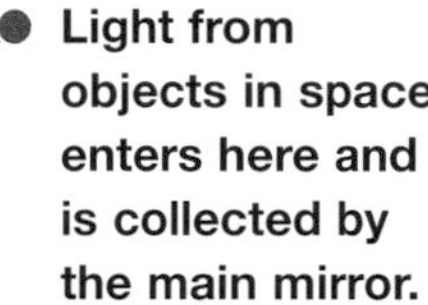

Light from objects in space enters here and is collected by the main mirror.

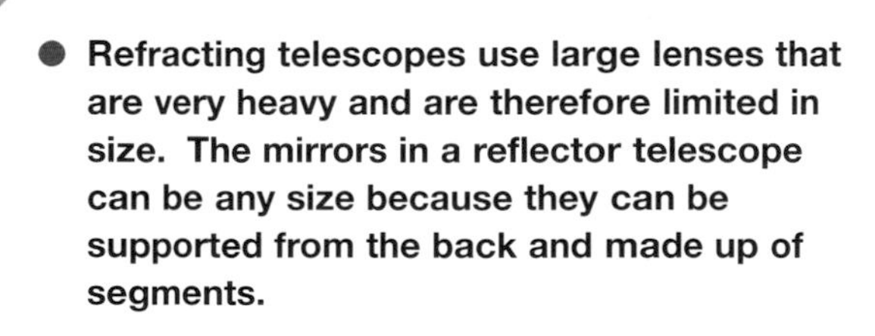

- **Refracting telescopes use large lenses that are very heavy and are therefore limited in size. The mirrors in a reflector telescope can be any size because they can be supported from the back and made up of segments.**

- **Refracting telescopes use thick lenses that absorb some of the light from faint objects. Lenses also focus different parts of the light spectrum at different points, which can blur images and make them difficult to view. As reflecting telescopes do not use lenses, they do not distort light in the same way.**

Modern telescopes

Astronomy today relies on computers as much as on telescopes. Data collected by telescopes and satellite dishes worldwide is processed by computers to produce images that can be stored and studied at leisure. Telescopes themselves are becoming bigger and therefore more powerful. The twin Keck telescopes, built on the 4,200-metre high summit of an extinct Hawaiian volcano, are the largest light-detecting telescopes in the world. The telescopes themselves are eight storeys tall, with mirrors ten metres wide!

Modern Astronomy

Telescopes have advanced a great deal in the four hundred years since their invention. There is now even a giant telescope in orbit around Earth. The Hubble Space Telescope has given astronomers incredible views of stars as distant as 13 billion light years away and changed the way we think about the Universe. Modern technology also allows us to "see" much more than just light.

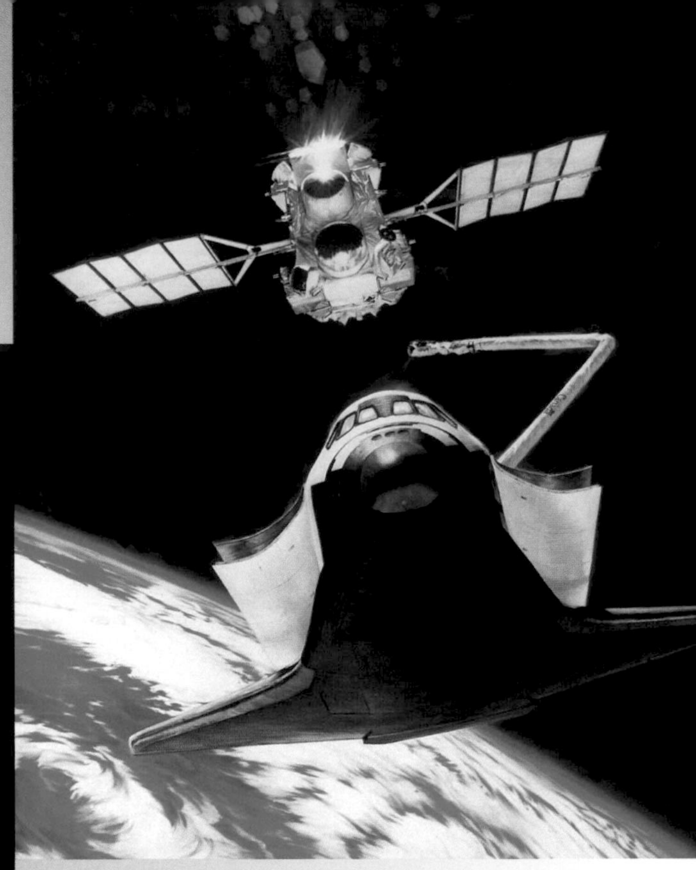

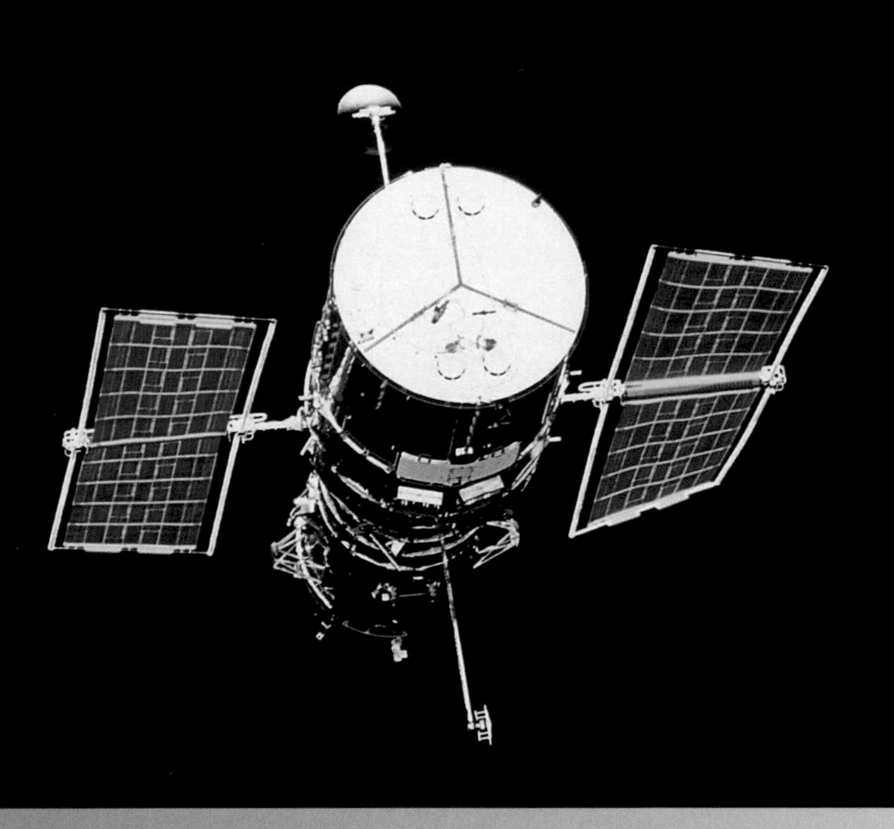

- There are all sorts of different satellites orbiting Earth. Many keep a 24-hour eye on space, viewing it in different wavelengths. The picture above shows the Space Shuttle deploying the Gamma Ray Observatory, which observes space in gamma ray wavelengths.

- Soon after it was launched, astronomers found that Hubble's main mirror was slightly the wrong shape, and the pictures were out of focus. In 1993, a servicing mission was sent to Hubble, and the telescope was given a corrective contact lens to improve its vision!

Hubble

The Hubble Space Telescope (above) was a dream come true for astronomers. Launched in 1990 after nearly half a century of planning, it gave a crystal clear view of space. Orbiting at 610 kilometres (380 miles) above the Earth's surface, its view is not blocked by the planet's turbulent atmosphere. Hubble has many different cameras, including infrared and faint-object cameras that can detect things that cannot be seen with the naked eye.

Astronomy in space

Most objects in space give off energy in the form of electromagnetic radiation. This radiation can be in many different forms and often travels millions of light years before it reaches Earth. However, most forms of radiation are absorbed by the Earth's atmosphere and so cannot be seen from the planet's surface. Because of this, these types of radiation are best studied from space. There are many different satellites orbiting the Earth, allowing us to study radiation before it reaches Earth's atmosphere.

Unusual astronomy

Amateur astronomers are usually only able to study visible light in the Universe. However, technology today allows us to “see” much more than this. Light is just one part of the electromagnetic spectrum. By studying different types of radiation, we can learn a great deal more about the Universe.

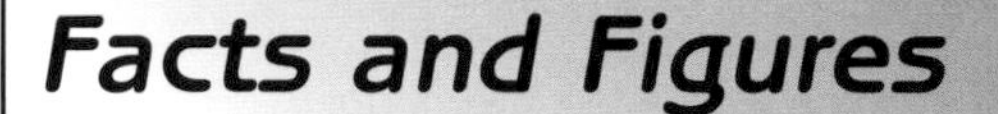

1800: William Herschel discovers infrared radiation when he splits sunlight through a prism.

1814: Joseph von Fraunhofer is the first to study the Sun’s spectrum using a spectrometer.

1932: Karl Jansky detects the first radio waves coming from space.

1970: The Uhuru satellite is launched to map the sky at X-ray wavelengths.

1983: IRAS is launched. It is the first infrared astronomy satellite.

1990: The Hubble Space Telescope is launched.

- **Radio astronomy**
  Radio telescopes collect the faint radio signals given out by objects in space. These telescopes are often large satellite dishes and are so accurate that they can track down molecules in the space between stars! This radio image is of volcanoes on Venus.

- **Infrared astronomy**
  Everything in the Universe that is cooler than normal stars emits infrared radiation. By looking at the Universe through an infrared telescope, we can see things that are not visible with an optical telescope.

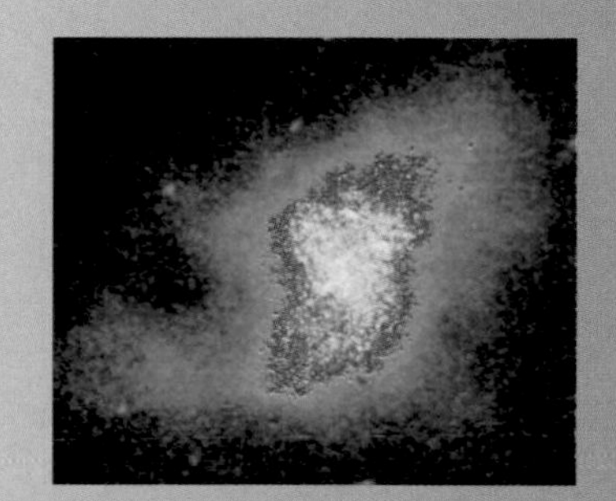

- **Ultraviolet and X-ray astronomy**
  These are radiation sources of a higher energy level than infrared and are best observed by telescopes orbiting Earth’s atmosphere. Ultraviolet astronomy (right) is used to track down the hottest stars. X-rays are emitted by even hotter objects, such as gas clouds around black holes.

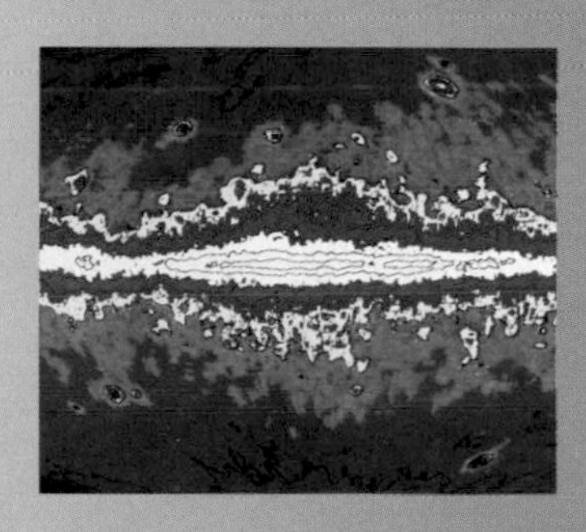

- **Gamma ray astronomy**
  Radiation of the highest energy levels is called gamma ray. No star or gas cloud could ever reach temperatures high enough to generate gamma rays. Instead, they are produced by colliding radioactive particles in space.

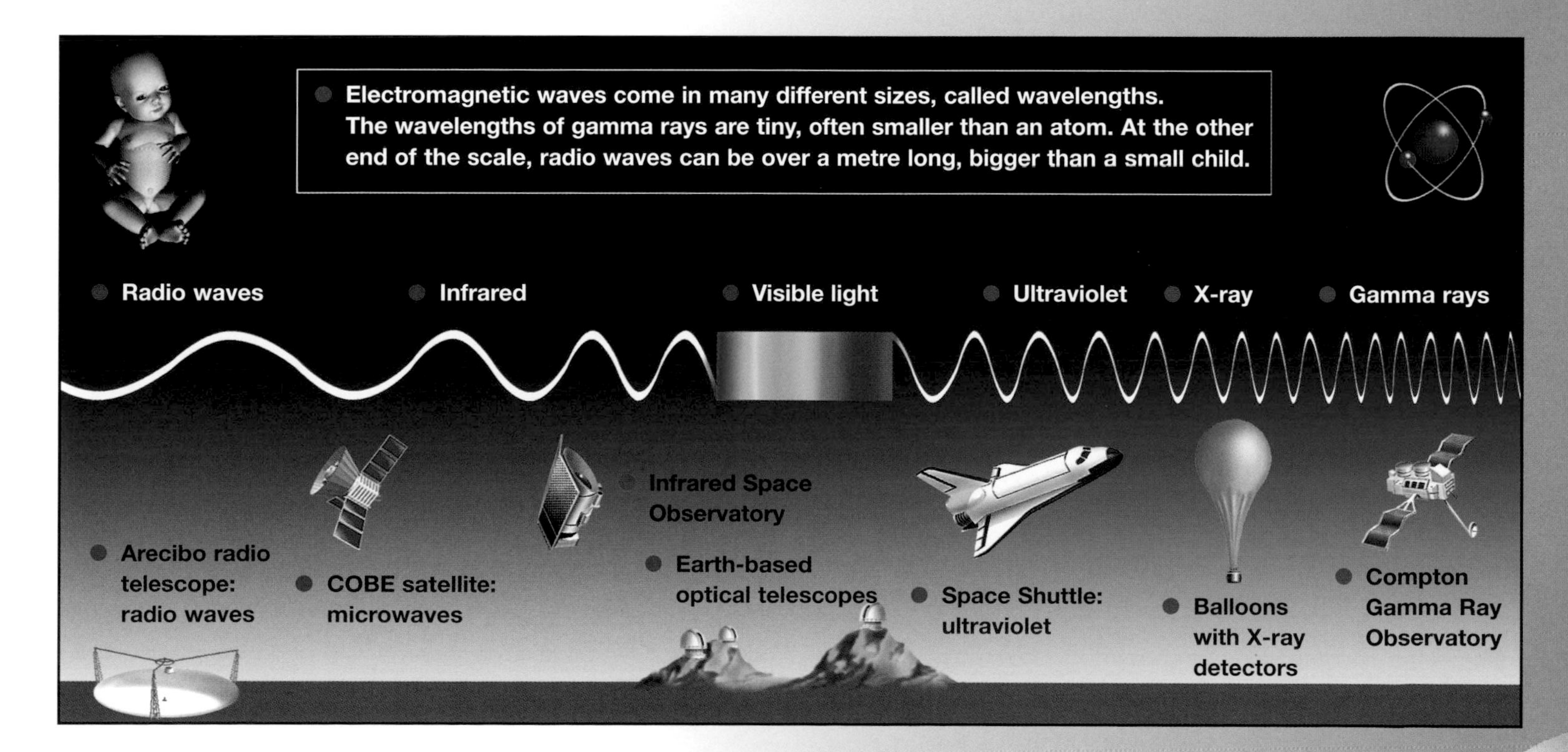

Into Space

Of all the marvels and achievements of the human race, the most incredible may well be our ability to leave our planet and travel into space. The drive for this came from two of the world's major superpowers, but it was not so much a desire for technological advancement as a battle for political supremacy. Nevertheless, Sputnik's bold voyage into space in 1957 remains one of the most important milestones of science.

The Space Age

The Space Age began with the launch of the first ever artificial satellite in 1957. Sputnik 1 was launched by the Union of Soviet Socialist Republics (the USSR, now divided into several countries, including Russia) and was the first human-made object to orbit the Earth. It did very little other than transmit a simple radio signal, but it marked a new stage in the history of humankind. This achievement for the USSR came as a surprise to the USA, which had previously announced its plans to be the first country to launch a satellite into space. A ferociously competitive war of supremacy began between the two superpowers. As a result, vast amounts of money were invested in space exploration.

Life in space

Less than a month after they launched Sputnik 1 into space, the Soviets claimed another major achievement by sending the first living creature into orbit. On 3 November 1957, Sputnik 2 blasted upwards from the Earth's surface containing a dog called Laika. She survived the launch but died when her supply of oxygen ran out in orbit. Three years later, two more dogs, Belka and Strelka, became the first creatures to survive the journey into space and re-entry, travelling in Sputnik 5.

- **Belka means "squirrel", and Strelka means "little arrow" in Russian. Strelka eventually gave birth to puppies, one of which was given to the President of the United States, John F Kennedy.**

- **Laika, shown left, travelled into space in a specially designed spacecraft. After Laika, Belka and Strelka, at least ten other dogs were sent into space by the Soviet Union. Five of them died.**

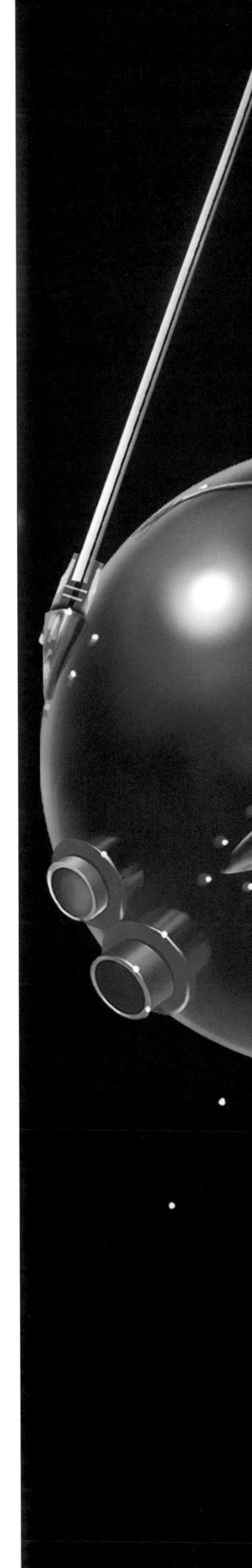

Many frightened people all over the world believed that Sputnik 1, shown below, was spying on their activities from space. However, all that it did was transmit a simple tracking signal for 21 days.

Man in space

Once more the USSR broke all records by sending the first man into space. On 12 April 1961, Yuri Alekseyevich Gagarin (shown below) was launched into orbit by a Vostok rocket. He completed one orbit of our planet before travelling safely back to Earth. The American crewed space programme was close behind, sending John Glenn into orbit in 1962. The first human being to leave the confines of a spacecraft and walk in space was the Soviet cosmonaut Alexei Leonov. He crawled through the airlock of Voskhod 2 in 1965 and was so overwhelmed by the view that he shouted out the first words he could think of: "The Earth is round!"

American and European men and women who go into space are called astronauts. People sent into space by the Soviet Union, and now by Russia, are called cosmonauts.

Facts and Figures

1957: The Soviet Union launches Laika, a dog, in Sputnik 2. She is the first living creature in space.

1960: Two more Soviet dogs, Belka and Strelka, become the first living creatures to survive the trip into space and back.

1961: Yuri Gagarin becomes the first person in space, flying in a Vostok spaceship.

1965: Alexei Leonov becomes the first person to walk in space. He spends 20 minutes outside the Soviet spacecraft Voskhod 2.

Rockets

If Ptolemy, Copernicus and Galileo were alive today, they would be amazed by how fast the human race has advanced in its exploration of the skies. No longer are astronomers restricted to peering through Earth's murky, temperamental atmosphere to view the planets and the stars. Space probes have visited every planet except one. Humankind has lived in space and even walked on the Moon. Rocketry has made all this possible.

Rockets

The ability to leave our planet and venture into the dark void of the Solar System is considered by many to be the most important technological advance in the history of the human race. It was made possible by the development of rocketry. Rockets, powered by gunpowder, were first used by the Chinese over 1,000 years ago. Rockets today work in the same way but use liquid fuel instead. Oxygen is needed in order for engines to work, and therefore the only engine that can work in the vacuum of space is a rocket, which carries its own oxygen.

Facts and Figures

900: The Chinese make solid rocket fuel out of gunpowder.

1895: Konstantin Tsiolkovsky suggests that a rocket can be made to work in a vacuum.

1926: Robert Goddard is the first person to launch a rocket powered by liquid fuel.

1944: German scientists develop the V2, the first ballistic missile to be powered by rocket technology.

- **The Mercury–Atlas launcher, shown left, was used in the USA between 1962 and 1963.**

- **The Vostock rockets, such as the one on the right, were designed in the Soviet Union. They put the first man in space when Yuri Gagarin travelled into orbit using a Vostock rocket in 1961.**

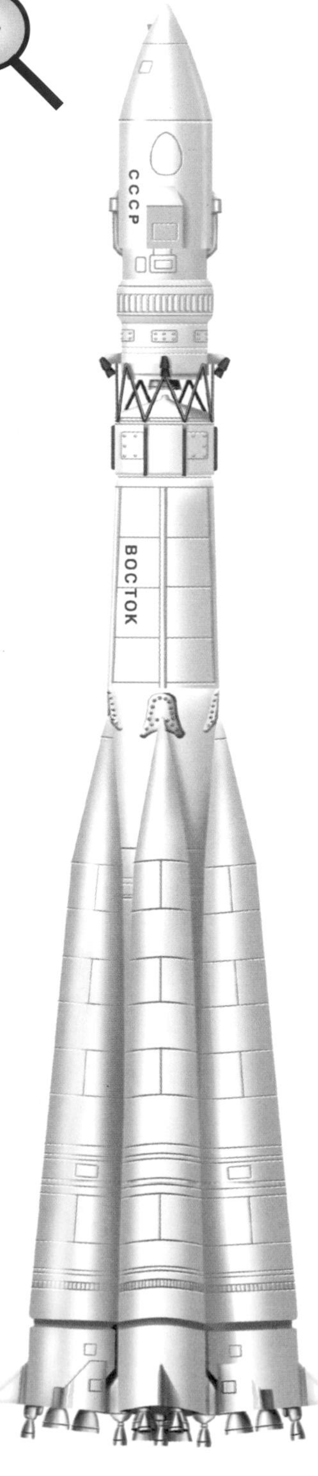

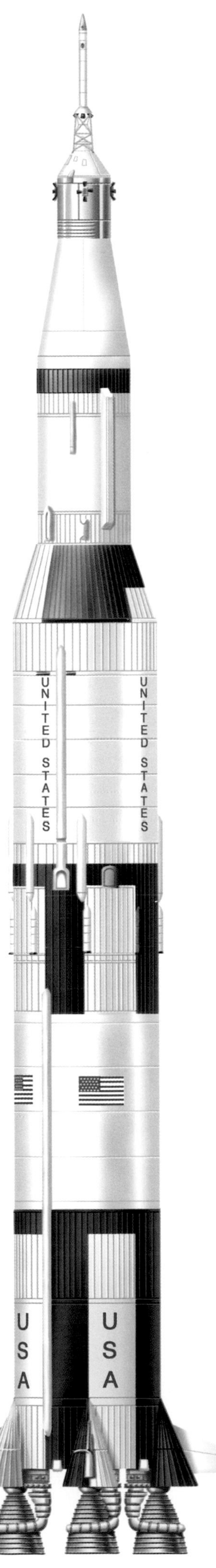

- A spacecraft must be able to fly faster than 30,000 km/h (18,600 mph) to escape the pull of Earth's gravity. If it travels slower than this, it will be pulled back to the ground.

- Big rockets are needed to carry the fuel required to blast into space. But the further a rocket travels from Earth, the weaker gravity becomes, so less power is needed. Because of this, rockets are usually made up of sections. As the fuel in a section is used up, the whole section is jettisoned, making the spacecraft lighter.

- The Saturn V rocket, left, was used in the USA between 1968 and 1972. It carried the first astronauts to the Moon in 1969. It can be seen taking off in the picture on the far left.

How rockets work

We are all held on Earth by the force of gravity. If there was no such thing as gravity, we would all be able to float into space at will, and life on the planet would certainly be extremely difficult! As it is, a vehicle wanting to fly into space must first create enough upward thrust to escape the clutches of gravity. A rocket does this by burning vast quantities of fuel and releasing the exhaust gases through a nozzle in the bottom of its engines (shown right). This produces enough force to push the rocket skywards.

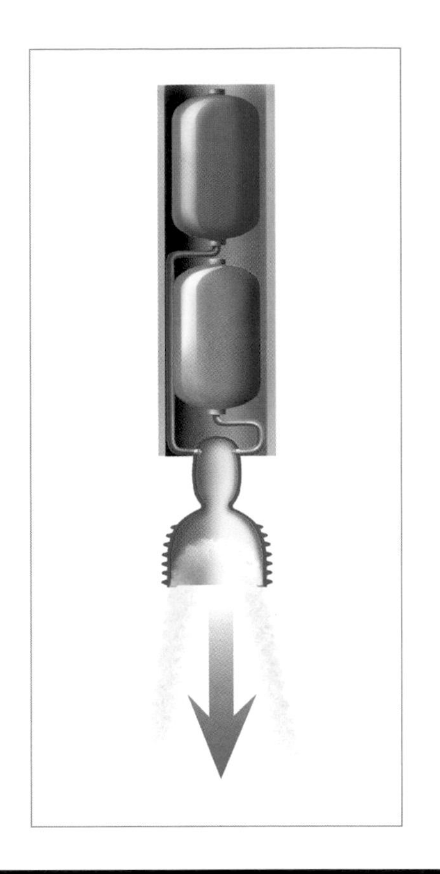

Ground control

All missions into space need support from scientists and engineers on the ground. There are many different factors concerned in a space launch. The astronauts alone would not be able to control all of them at once. At the National Aeronautics and Space Administration (NASA) Mission Control Centre in Houston, USA, shifts of flight controllers check everything on the spacecraft to make sure the mission is going to plan. This includes making sure the craft is on the right course, checking communication links and ensuring there is enough oxygen and fuel to complete the flight.

Space Shuttles

Rockets are essential for space travel, providing the thrust needed to escape the pull of Earth's gravity. However, rocket launches are incredibly expensive, especially as most of the vehicle is destroyed in the launch procedure! Engineers designed the Space Transport System, often called the Space Shuttle, to be re-used mission after mission and cut costs.

Anatomy of a shuttle

The Space Transport System (STS), better known as the Space Shuttle, is made up of three main parts. The largest section of the Shuttle is the main external fuel tank. This is taller than a 15-storey building and is the only section of the Shuttle that cannot be re-used. It is jettisoned after eight minutes and burns up as it re-enters the Earth's atmosphere. The solid-fuel rocket boosters propel the rocket to a height of 45 kilometres (28 miles) before they are dropped into the Pacific Ocean to be picked up and re-used. The orbiter is the main section of the Shuttle, housing the crew and the payload bay, where the Shuttle's cargo is kept. The Shuttle only carries one orbiter at a time, but there are four to choose from, and each has a payload bay that can hold five cars end to end.

Lift-off!

A Space Shuttle launch is an incredible sight. Plumes of fire blast from the rocket boosters as the Shuttle lifts skywards. The Shuttle's weight on Earth is more than two million kilograms (four million pounds), so the vehicle's journey into space starts quite slowly. It takes eight seconds to accelerate to 160 km/h (100 mph), but soon speeds up. After one minute, the shuttle has accelerated to an incredible 1,600 km/h (1,000 mph). These awesome speeds are needed if the craft is to leave Earth's atmosphere. To achieve orbit, the shuttle must be travelling at 30,000 km/h (18,600 mph). This is ten times faster than a bullet!

Re-entry

Re-entering the Earth's atmosphere is a dangerous procedure. The Shuttle begins to slow down half a world away from its destination—the runway at the Kennedy Space Center in Florida, USA. Because the Shuttle is travelling so quickly, it does not need engines to navigate its way back to Earth. Instead, it becomes a giant glider, using the air to slow it down as it heads for home. Tiny particles in the Earth's atmosphere collide with the Shuttle as it charges downwards, creating heat through friction. Temperatures on the Shuttle's wing tips and nose can reach a searing 1,700°C (3,000°F)!

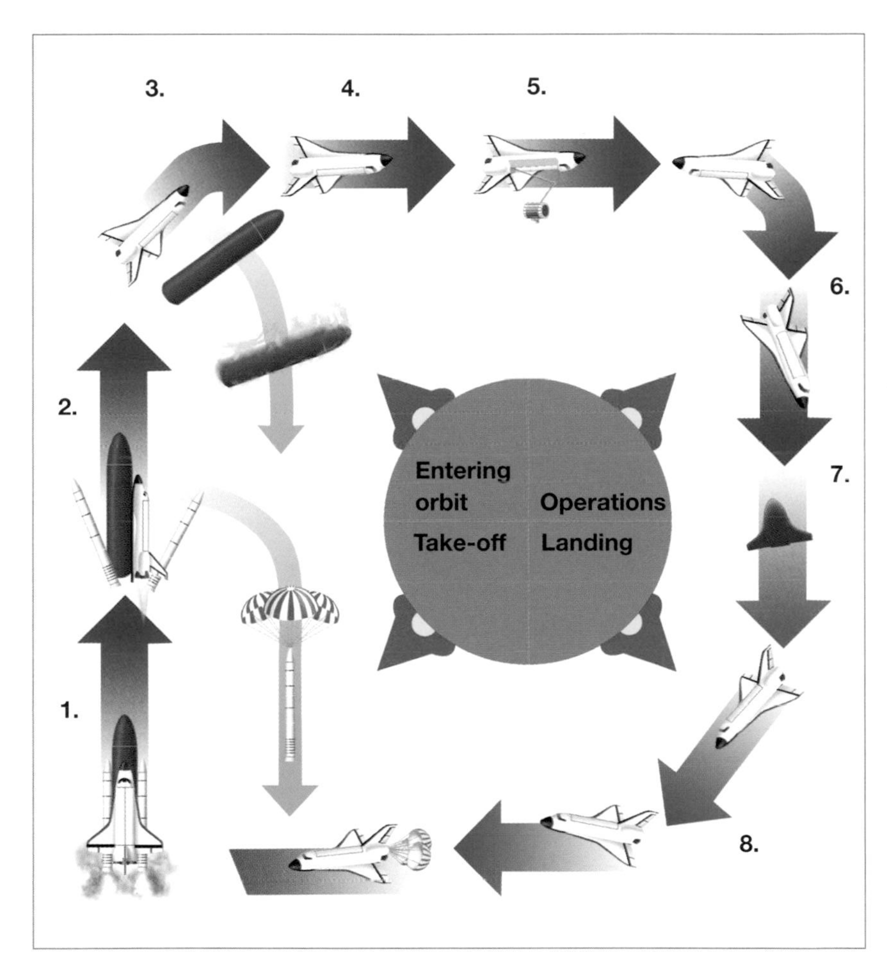

1. The Space Shuttle blasts off from the launch tower.

2. The fuel in the rocket boosters is used up and they are jettisoned. They fall back to Earth, using parachutes, to be recycled.

3. The external fuel tank is released. It burns up in the atmosphere.

4. The orbiter reaches a low-Earth orbit.

5. The orbiter may stay in space for over two weeks, doing a variety of tasks. These may include deploying satellites or carrying astronauts to repair equipment on the Hubble Space Telescope.

6. The orbiter positions itself ready for the return journey to Earth.

7. As the orbiter re-enters Earth's atmosphere, friction causes it to heat up.

8. The orbiter glides in and prepares for a high-speed landing on a 4.5 km (2.8 mi) runway. Parachutes help it slow down.

New Rockets

The rockets used today are not very different to those used in the Apollo programme over thirty years ago, and they are just as expensive to build. However, recently there has been a large demand for satellites, and because of this, scientists are looking for cheaper ways to send payloads into Earth's orbit. The Space Shuttle was a step forward in designing a re-useable spacecraft, but scientists are now coming up with ideas that could open up space to everybody!

The X-33

NASA's Single-Stage-to-Orbit Reusable Launch Vehicle is the first new spacecraft for over twenty years. It is currently a sized-down prototype of a new rocket called the Venturestar (right), which will be built if the X-33 is successful. Whereas the Space Shuttle jettisons its boosters and main fuel tank during its journey into orbit, the Venturestar will travel into space and back in one piece. The rocket boosters will be replaced by an engine called a linear aerospike, which is powerful enough to launch the spacecraft without the need for external rockets. The X-33 takes off vertically in the same way as the Space Shuttle and lands like a plane. However, its design will lower the price of putting a pound of payload into orbit by about 90%.

Other suggestions

The X-33 is only one of many recent attempts to design more efficient space rockets. The DC-XA, shown left, was a revolutionary new rocket made from pieces of equipment re-used from other vehicles. It could take off and land vertically but was discontinued after four test flights because of a crash. Another remarkable idea is the Roton, which is designed to make rockets lighter by removing the heavy technology needed to pump rocket fuel. Instead of mixing fuel and oxygen at high pressure, the rotor blades on the Roton spin the engine so that propellant is literally thrown into the combustion chamber. On the return to Earth, the blades provide lift in the same way as a helicopter's rotor blades.

- The Deep Space 1 probe is one of the most successful tests of ion technology. It has been running in space for longer than any other rocket in the space programme.

- Conventional rockets are very expensive and noisy to use. Ion propulsion engines use only 100g (3.5 oz) of xenon per day, which is released as a faint blue glow rather than an explosion.

Ion propulsion

Scientists have developed a new kind of rocket technology that is a great deal cheaper than using liquid or solid fuel. Ion propulsion engines work in the same way as conventional rocket engines, by expelling a force in one direction to help propel the spacecraft in the opposite direction. However, instead of using the exhaust gases from mixing liquid hydrogen and liquid oxygen, ion technology runs an electrical charge through a gas called xenon. This causes ions (basically charged atoms) to shoot out of the back of the spacecraft at a speed of over 100,000 km/h (62,000 mph), much faster than the gases expelled by conventional rockets.

- The X-33 has a lightweight titanium body that makes the craft much lighter than the Space Shuttle. Its weight on the ground is 123,800 kg (273,000 lb), of which 95,300 kg (210,000 lb) is fuel!

- With much cheaper rockets being developed, space travel will soon be available to many more people. A number of people have already taken a holiday in space, and others have made reservations to do so in the future.

- The xenon gas used in ion technology is the same gas that is used in camera flash-tubes and lighthouse bulbs.

Space Probes

While early telescopes were vital for showing astronomers the orbits of the planets, they were not able to provide much information about the planets themselves. As late as the 1950s, astronomers were making assumptions about the planets that could not be verified using telescopes—such as the possibility of vegetation on Mars and Venus. Only with the advent of space probes have we learnt more of the truth about the Solar System.

● **The Viking probes visited Mars in 1976 in search of life. They conducted numerous tests on Martian soil but found no evidence of any living organisms.**

● Different duties

Probes are normally designed to do one of three things when they reach their destination. Some fly by a planet at a distance of several hundred kilometres, taking pictures of the planet from this distance. Others are called orbiters and are able to gather more detailed information about a planet because they can orbit it for a longer period of time at a lower height. The probes that get closest to the planets are called landers. So far, probes have only landed on the Moon, Mars and Venus, although the Galileo spacecraft dropped a small probe into the gassy atmosphere of Jupiter.

● **The Mars Polar Lander**

● **The Mariner 10 probe flew by Mercury three times in order to photograph the planet's surface. It flew within 327 km (203 mi) of the planet but because of its flight path it could only photograph one half of Mercury.**

Landers

Landers are the only human-made things to have touched down on the surface of another planet in the Solar System. The Soviet-built Luna 2 was the first space probe to touch the Moon's surface, in a crash landing. Mars is the most visited of the worlds in our Solar System. More than ten probes have visited the red planet since 1964. Many probes have actually touched down on the surface to bring back incredible information about the planet's composition and environment. In 2004, NASA's Opportunity and Spirit rovers landed on Mars and transmitted detailed photographs of the ground features and began analyzing rocks and soil for evidence that water once existed on the planet's surface.

Facts and Figures

Important probes:

1973: Pioneer 10 was the first probe to cross the asteroid belt.

1974: Mariner 10 was the first and so far only probe to visit Mercury.

1976: Viking 1 and 2 were the first probes to search for life on Mars.

1986: Giotto was the first probe to photograph the nucleus of a passing comet.

1991: Galileo was the first probe to take close-up pictures of an asteroid (Gaspra).

The Galileo probe was not actually launched towards Jupiter, its target, but towards the Sun. The probe used the Sun's gravity to catapult it across the immense distances to Jupiter.

Beyond the planets

Four probes are hurtling out beyond the planets as you read this. NASA's Jupiter and Saturn probes, Pioneer 1 and 2 flew by their respective planets and kept on going deep into the far reaches of our Solar System. Each carries an engraved plate that is a message for any extraterrestrial beings who find them. The two Voyager probes, launched in 1977, carry gold-plated records containing sounds and images from Earth. These probes will continue to head out of the Solar System, measuring the extent of the Sun's solar wind. They will eventually run out of power but will keep drifting until they break up, collide with something or get picked up by aliens.

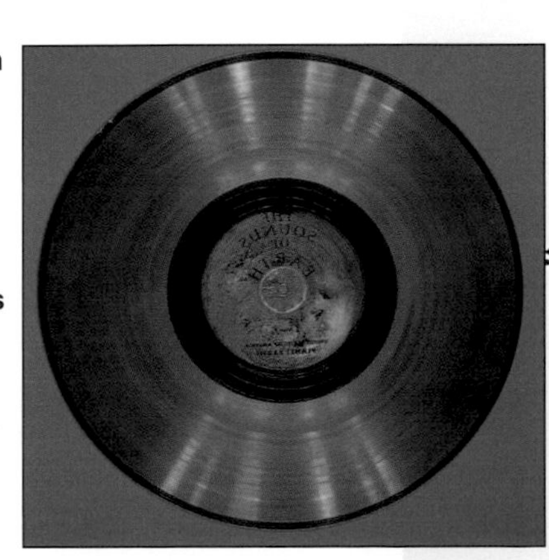

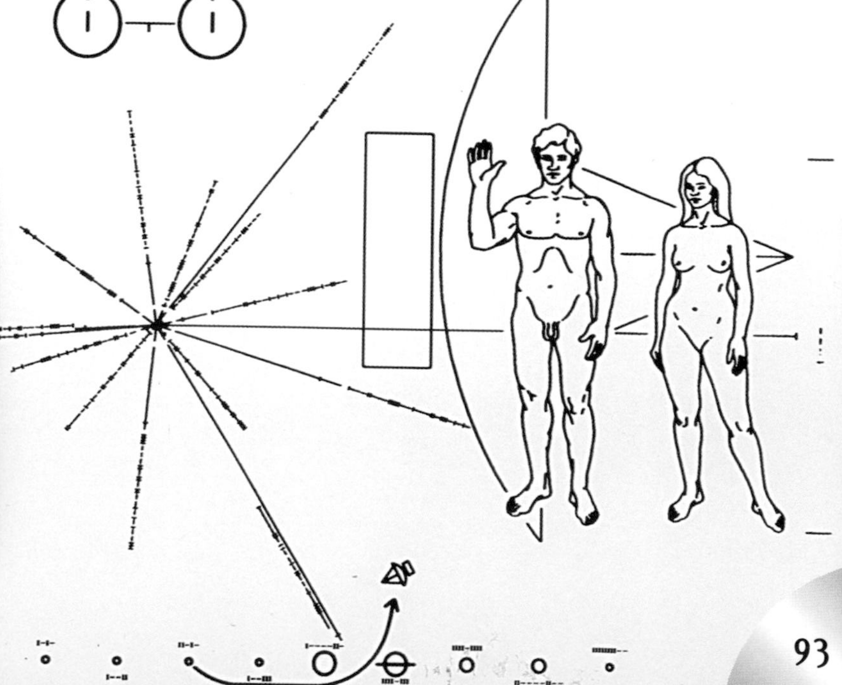

The engraved plate on the Pioneer probes shows a man and a woman in a greeting stance, and Earth's position in the Solar System. The gold disks on the two Voyager probes contain, amongst other things, music by Beethoven and greetings in over fifty languages.

The Apollo Missions

Apollo was the ancient Greek god of the Sun, but his name is now synonymous with another celestial object, the Moon. The Apollo space programme was set up by the USA in 1961. President John F Kennedy ambitiously claimed that human beings would set foot on the Moon by the end of the decade. The motives for this were as much political as they were scientific, but Neil Armstrong's "small step" will always be an incredible moment in space exploration.

Voyage to the Moon

A special rocket called the Saturn V was designed in the 1960s. It was able to carry the 52 tonnes of equipment needed for a manned mission to the Moon. It made its maiden flight in 1967 and successfully took three astronauts to the Moon two years later. The rocket carried a special piece of equipment called the Lunar Module. This specially built craft took Neil Armstrong and Buzz Aldrin to the Moon's surface. As well as providing their transport, the Lunar Module was the astronauts' home for the three days they spent on the Moon.

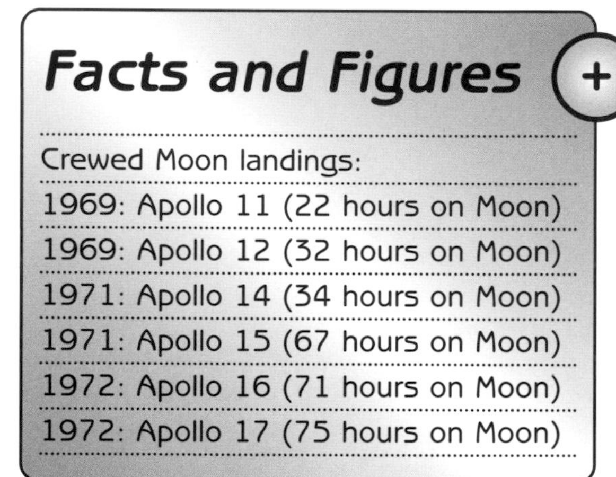

Facts and Figures

Crewed Moon landings:

1969: Apollo 11 (22 hours on Moon)
1969: Apollo 12 (32 hours on Moon)
1971: Apollo 14 (34 hours on Moon)
1971: Apollo 15 (67 hours on Moon)
1972: Apollo 16 (71 hours on Moon)
1972: Apollo 17 (75 hours on Moon)

Because there is no weather on the Moon, the astronauts' footprints will last for thousands of years!

Apollo astronauts

Altogether there have been six Apollo missions to the Moon, during which 12 astronauts have explored its conditions and composition. Thousands of photographs and 388 kilograms (176 pounds) of Moon-rock have been brought back to Earth for scientists to study in their quest to discover the origin and history of the Moon. Astronauts have also studied solar influences on the Moon and measured the amount of dust in the air. They have tested the soil to try to determine what the Moon is made from, and measured Moonquakes, which are slight movements in the Moon's crust.

The Lunar Rover

In later Apollo missions, the astronauts took the Lunar Rover with them. This was a battery-powered vehicle that enabled the astronauts to explore much more of the Moon than their predecessors had been able to do on foot. The Moon Buggy, as it is often called, had a television camera and a satellite dish so that images could be sent back to Earth. Its tyres were solid to avoid damage from the rough surface of the Moon. The Lunar Rover could be folded up and stored away so that it did not take up much room on the journey.

- **Conditions on the Moon proved very strange for the astronauts. Gravity is only one sixth of that on Earth, which means that astronauts can jump several metres effortlessly.**
- **There is no atmosphere on the Moon, which means that sound cannot be carried even over a tiny distance. Radios have to be used to communicate over a few centimetres!**
- **The Moon Buggy was steered by a small hand-control instead of a wheel. It carried tool kits and sample bags.**
- **The Lunar Rover was powered by a battery. Its top speed was just under 20 km/h (12 mph).**

Space Stations

Only the most optimistic of science-fiction fans at the beginning of the twentieth century believed that one day human beings could live in space. However, scientists have designed space stations that can sustain human life in the hostile conditions of space. Life support systems that provide air and water mean that astronauts and cosmonauts can live in space for periods longer than a year.

The ISS

In 1998, the first module of the International Space Station was launched. Many different countries are involved in this colossal task. The space station was originally scheduled to be finished in 2006, but there have been major delays. When it is complete, the space station will be the size of a football field and the brightest object in the night sky after Venus.

- **The International Space Station is powered by over 4,000 square metres of solar panels.**
- **The International Space Station will house a crew of seven. They will live in a special habitation module with a gym, a galley and medical facilities.**

Facts and Figures

Space station launch dates:

1971: Salyut 1 (Soviet Union)

1973: Skylab (USA)

1983: Spacelab (Europe)

1986: Mir (Russia)

1998: International Space Station

Mir

The Mir space station, shown above, went into orbit around the Earth in 1986. It orbited the Earth every 90 minutes at a height of 320 kilometres (200 miles). Despite the cramped conditions inside Mir, some cosmonauts remained on board for over a year. After the collapse of the Soviet Union in 1991, Russia took over the operation of Mir. Russian cosmonauts repaired major breakdowns on the ageing station until, in 2001, Russia took Mir out of orbit and sent it plummeting to Earth.

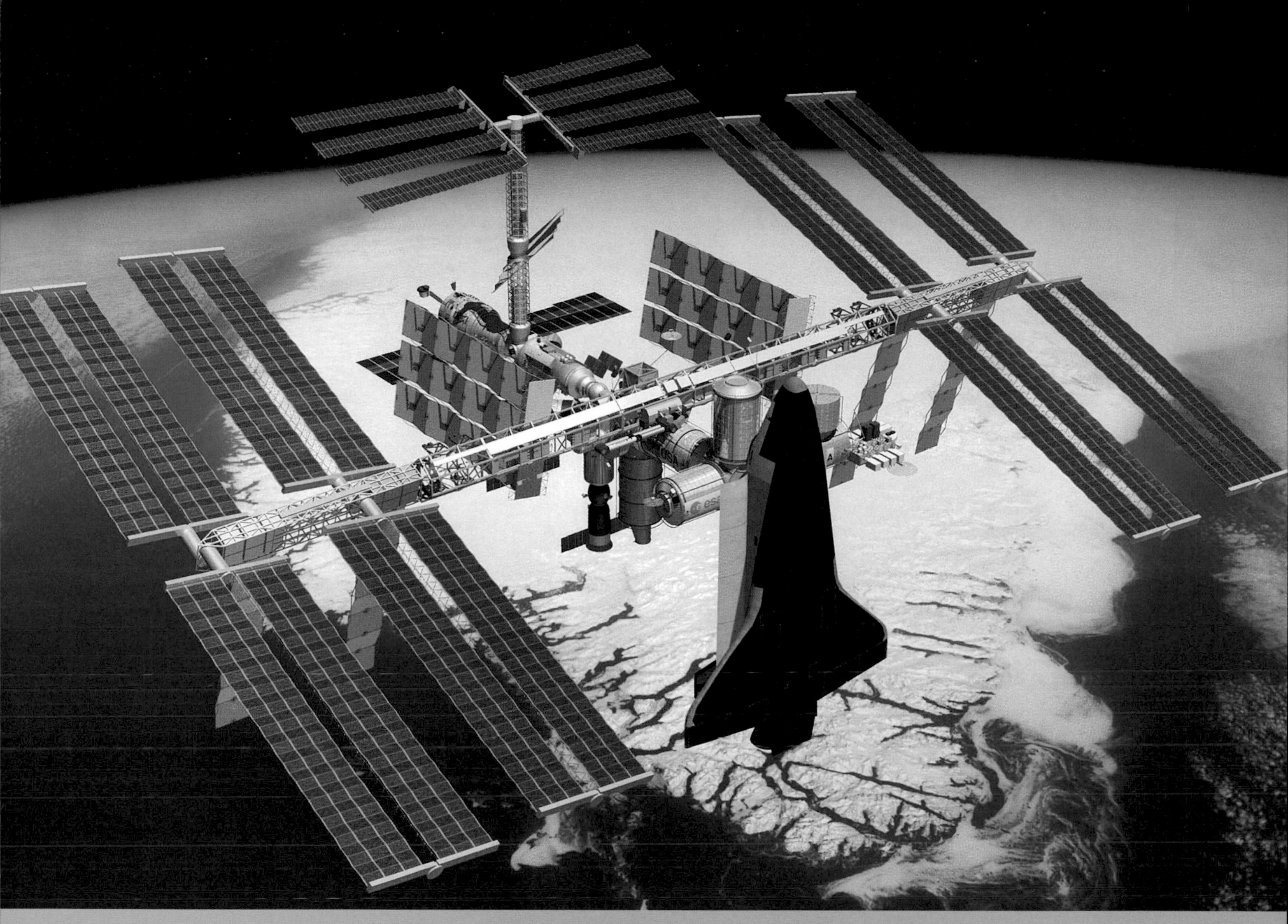

The International Space Station will be supplied by the Space Shuttle, which will dock in order to provide food and equipment.

The International Space Station has a wingspan of 110 m (120 yd), is over 80 m (88 yd) long and weighs 500 tonnes.

Conditions in space

Conditions in space can be very strange. There is almost no gravity inside a space station, which means that astronauts can float in mid-air and lift heavy objects effortlessly. There is no up or down in space, and astronauts can eat or sleep on the walls, or even on the ceiling! This lack of gravity can be troublesome, however, as scientists in space stations have to strap themselves to the walls when they are working to stop themselves from floating away. Because muscles do not have to fight against gravity in space, they can waste away, which means that astronauts must exercise every day. There is also no proper day and night on a space station. For example, on Mir the sun rises and sets every 90 minutes.

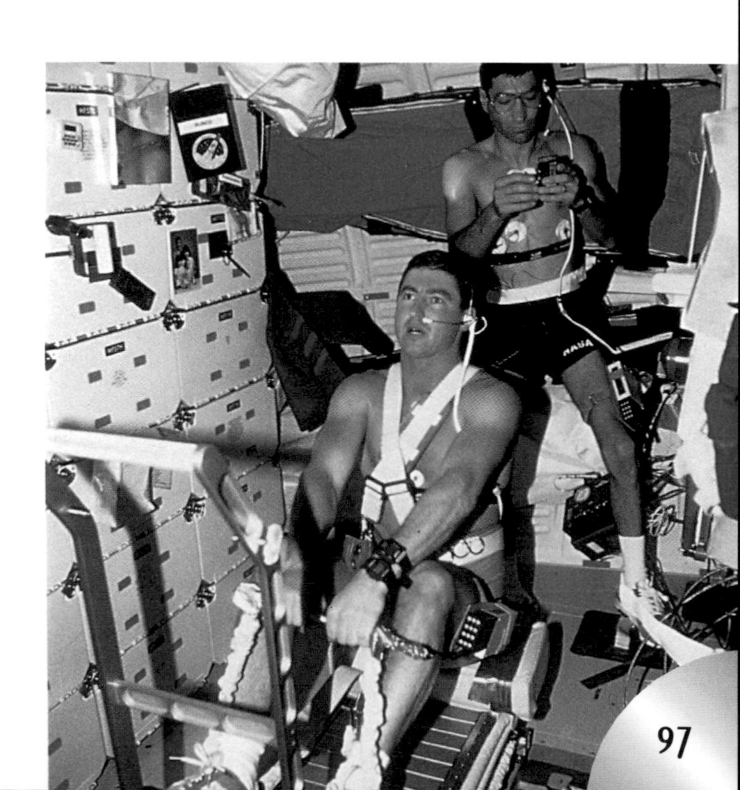

Space Science

Until humankind discovered the technology that allowed it to work off-planet, science was always governed by gravity. The first space station, Salyut 1, was launched in 1971. Since then scientists have had an incredible new field of physics to explore—that of microgravity. Space stations are offering ever more advanced facilities in which to explore how things work differently when there is almost no gravity.

- Any object with mass has gravity. Large objects such as planets and stars have enormous gravitational pulls, but even smaller things such as you and I create tiny gravitational forces. When you throw a ball in the air on Earth, it will always come back down because the force of gravity pulls it towards Earth's core. In a space station, there is no mass great enough to produce a significant gravitational pull.

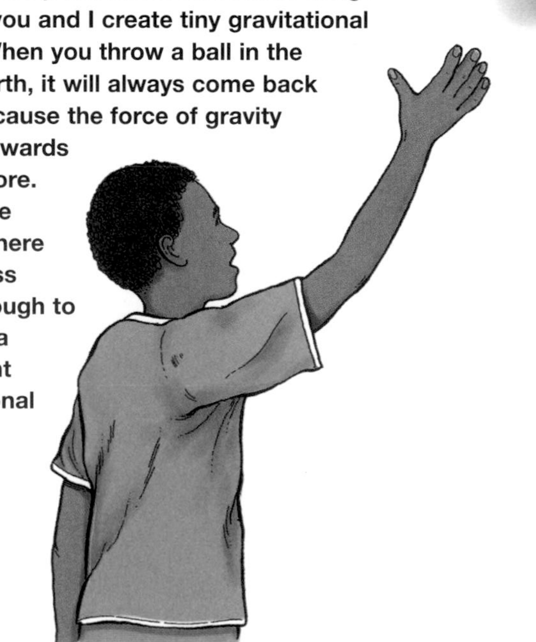

Why space?

Gravity is one of the fundamental forces in the Universe. It is what attracts objects with mass to one another and is why a dropped egg will always be pulled towards the ground rather than float conveniently in mid-air in one piece! Everything on Earth is influenced by gravity, from the way our bodies work to the way that plants grow, from the design of rockets to the development of crystals. Because gravity cannot be changed on Earth, scientists instead have ventured into space to work in environments where gravity is so low it is almost non-existent. This is called microgravity.

Skylab was the first US space station. It ran many tests to see how the human body behaved in microgravity.

Microgravity

Working in space can help scientists explain how different things are affected by gravity. They do this by noting how things respond differently with no gravitational pull. The European Space Agency's Spacelab was designed with two pressurized laboratories where microgravity experiments could be carried out. Special racks held hundreds of different kinds of cells and organisms including bacteria, lentil seedlings and shrimp eggs! Scientists studied these organisms to see if they behaved differently in space. Spacelab also studied how the human body is affected when it is in space for a long time. Tests were run on the heart, lungs and digestion.

Technology from space

Living organisms are not the only things that can be studied in space. Scientists have also studied combustion in microgravity to help design more efficient jet engines, and crystal growth to help build better semiconductors for computers. We have all benefited from technology that was designed for use in space. Microchips that are found in everything from digital watches to computers were first developed so that lots of equipment could fit into a small spacecraft. Technologies such as keyhole surgery and solar power have advanced because of science from the space programme. Many household items have also come about because of space technology, such as air-tight cans, kitchen foil and Velcro.

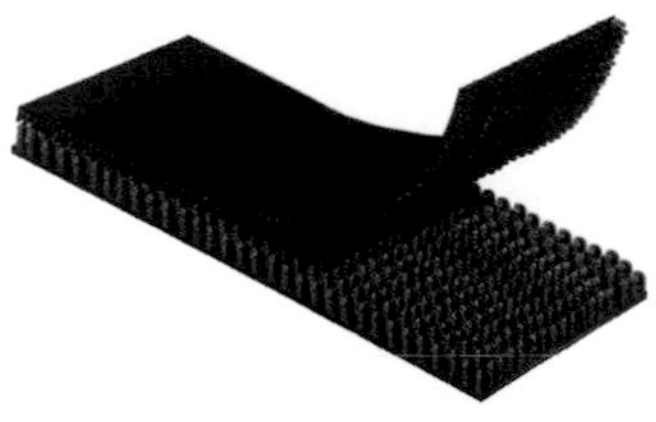

Working in Space

Imagine the feeling of stepping out of the safety of your spacecraft into the emptiness of space. The vast, blue globe of Earth lies 320 kilometres (200 miles) below you, and perhaps you can even see your home continent beyond the wispy clouds. One step away and you will drift into the microgravity of Earth's orbit, but one wrong move and you might find yourself floating into the infinite depths of space.

Conditions in space

Space may seem to be a calm and quiet place when viewed from Earth, but it can be deadly. If humans ventured unprotected into the airless vacuum of space, they would die almost instantly. With no pressure, the gases in their blood would separate as though it was boiling. With no atmosphere to filter out the Sun's harmful ultraviolet rays, the radiation would fry them in seconds!

Moving in space

Astronauts have to work in space. They often have to repair components of satellites or space stations. Movement in space can be difficult and this can hamper these tasks. Microgravity means that an astronaut is in danger of floating away in mid-job, or losing a vital tool into outer space. Handles and special footholders, into which feet can be locked, help astronauts move around in space. When they have to fly further away from their Shuttles, astronauts can use special backpacks that have rockets built into them. Called Manned Manœuvring Units, these backpacks are like floating armchairs and are directed by a hand-control like that for a computer game.

Spacesuits

The Extra-Vehicular Activity (EVA) spacesuit acts like a miniature spacecraft. It provides everything that an astronaut needs to survive for short periods in space, including oxygen to breathe, water to drink, heating and cooling, communication devices and toilet facilities! These suits have been developed to withstand the extreme conditions of space. Layers of different materials protect the astronaut from tiny particles of space dust that travel at hundreds of thousands of kilometres per hour. The suit maintains a constant pressure by surrounding the body with a kind of balloon. This balloon is full of air, which presses against the body in the same way as Earth's atmosphere.

- Because there is no gravity in space, astronauts can move giant pieces of equipment that would weigh many tonnes back on Earth.

- This spacesuit has a shiny, gold-coated visor to protect the astronaut's eyes from the blinding light of the Sun.

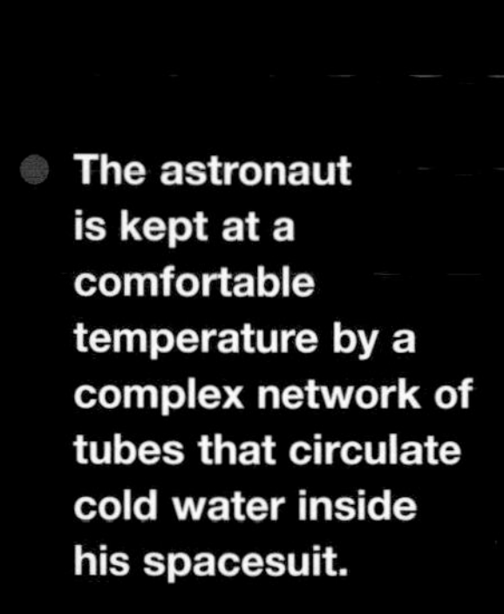

- The astronaut is kept at a comfortable temperature by a complex network of tubes that circulate cold water inside his spacesuit.

- Designing spacesuits is a very complex procedure. They have to be strong enough to resist the harsh conditions of space but flexible enough to allow the astronaut to perform delicate manœuvres. The astronaut on the right is making adjustments to the fragile equipment on the Hubble Space Telescope. His spacesuit gives him complete control and protection.

Satellites

Anything that orbits a planet is called a satellite. Earth has had its own satellite for billions of years—the Moon. However, the last fifty years have seen hundreds of human-made satellites launched into orbit around Earth, each transmitting a cacophony of radio signals. Modern satellites are incredibly powerful. Some could even read the words in a book you were studying in your own garden!

Satellite uses

Satellites are vitally important to many areas of modern life. You probably use many different satellites every day without even realizing it. You may watch satellite television while you eat your breakfast, or use the internet at school. The chances are that either of these activities involves signals being bounced off an overhead satellite. If you speak to relatives in another country by telephone, it is highly likely that your voice is travelling through space at the speed of light! If you are planning a day out, meteorology satellites bring you up-to-date information on approaching weather systems.

Facts and Figures

Launch dates:

4 October 1957: Sputnik 1, the first satellite

1 February 1958: Explorer 1, the first US satellite

7 August 1959: Explorer 7, the first satellite to study climate

1 April 1960: TIROS 1, the first weather satellite

13 April 1960: Transit 1B, the first navigation satellite

6 April 1965: Early Bird, the first communications satellite

Satellite orbits

Different kinds of satellites are programmed to follow different orbits. There are four main orbits used for satellites, each suitable for satellites with different purposes. A satellite in a geostationary orbit takes the same time to orbit the Earth as the Earth takes to spin, therefore always remaining over the same place. This kind of orbit is usually used for communication satellites. Low-Earth orbits are often as low as 250 km (155 mi) above the planet, and are most often used by spy satellites. Polar-orbit satellites orbit at a slightly higher 800 km (590 mi), while highly-elliptical-orbit satellites orbit Earth at low altitudes but pass far beyond Earth when they are at their most distant.

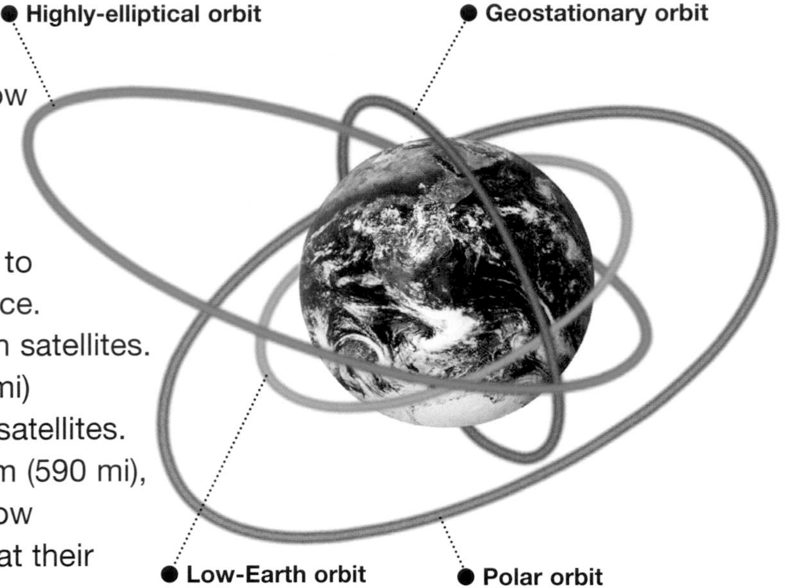

Military satellites

- Some satellites detect the radiation given off by nuclear explosions. This often deters countries from testing their nuclear arsenals.

Over half the satellites sent into space by Russia and the USA are for military purposes. There are many different kinds of military satellites with different purposes—from taking pictures of enemy launch-sites to eavesdropping on telephone calls. The first military satellites were launched in the 1950s by the USA and took pictures of enemy territory before returning home to have their film developed. Modern technology allows satellites to photograph the Earth electronically, sending back information digitally and so never running out of film. Amazingly, satellites can photograph things as small as words on a newspaper from space! The picture below shows a satellite image of Washington DC, USA.

- Superpowers like the USA and Russia use satellites to warn them of enemy missile launches. The American Space Based Infrared Satellite System detects the heat given off by ballistic missile launches.

- Some satellites detect dangerous weather systems such as hurricanes, which can be deadly if people are not warned. Satellites can inform people of the danger in time.

Space Waste

Just as humankind has discarded many unwanted materials in the ground and the ocean, there is now a great deal of pollution in space. There are millions of pieces of space debris orbiting the Earth, ranging from football-sized remains of old satellites to minuscule pieces of dust. Even something as small as a fleck of paint can be fatal if it hits a fast-moving spacecraft.

Space remains

Space debris is mainly the remains of spacecraft that have been discarded, or which have exploded in orbit around the Earth. Rubbish is created at many stages during the launch of a spacecraft or satellite. Boosters are often discarded when empty, and the nose cone of a rocket is always thrown away when a satellite is released into space. Obsolete satellites are often left in space and could orbit for thousands of years. As more and more satellites are launched into space, orbits are fast becoming littered with unwanted space debris, and the danger of collisions is increasing.

If pollution in space continues, before long we may no longer be able to practise astronomy from Earth because the skies will be full of junk! The picture to the left shows one artist's impression of how clogged up Earth's orbit has become.

● Scientists are working on ways to get rid of space debris. The artwork on the left shows one suggestion. A specially designed satellite orbits Earth and destroys loose pieces of debris with a laser beam. The laser would have to be powerful enough to obliterate the junk completely, as even a collision between a spacecraft and a speck of paint can be disastrous.

● Astronomers estimate that there could be up to 10,000 items of space debris larger than 10 cm (4 in) orbiting the Earth. This artist's impression shows how dangerous it can be for craft travelling into space.

● Space cleaners

Because of the dangers posed to spacecraft, scientists are beginning to think of ways of controlling the amount of space debris in Earth's orbit. Radars look out for bigger pieces of junk and warn spacecraft in advance. Spacecraft and space stations are beginning to be coated with a special ceramic fibre that can absorb the energy of a collision. As for the vast amounts of junk already out there, some scientists have suggested cleaning robots, which could travel through Earth's orbit collecting pieces of space junk and taking them back to special space stations to be recycled.

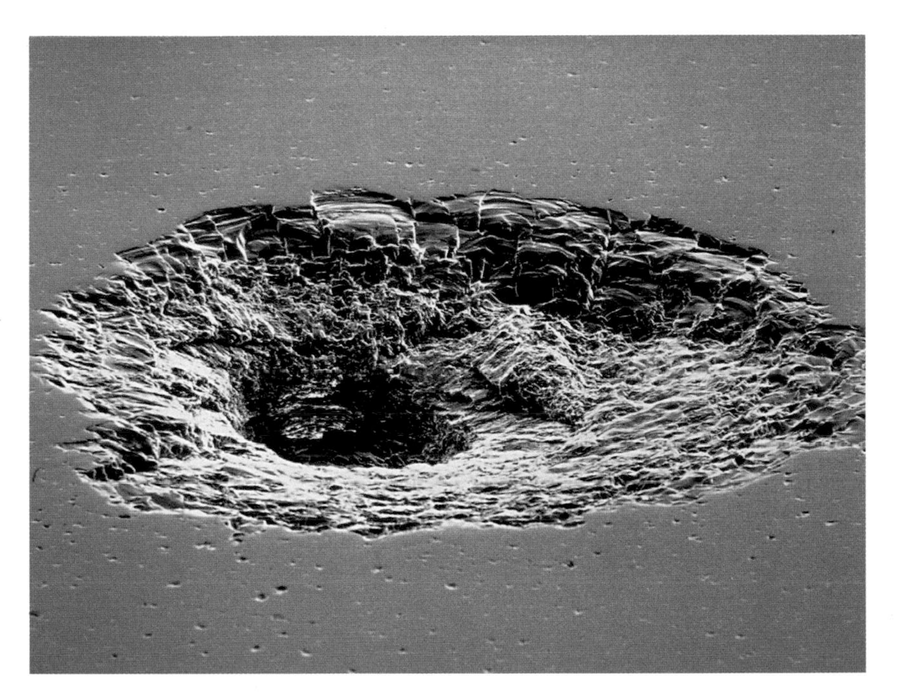

● A minuscule speck of paint less than 1 mm (0.04 in) big made a hole of 0.5cm (0.2in) in the window of a Shuttle when it collided in 1993.

● Dangers in space

Most of the material orbiting Earth is now classed as space debris because it has no use. Scientists estimate that there could be more than 40 million pieces of space junk, much of which is larger than 10 centimetres (4 in) in size. Spacecraft such as the Space Shuttle move at incredibly fast speeds, and scientists worry that if these craft collide with pieces of space debris, the results could be disastrous. A fleck of paint travelling in the opposite direction to a Space Shuttle could impact at speeds of 40,000 km/h (25,000 mph) or greater. An impact of that magnitude could easily smash the orbiter's window, depressurizing the cabin and killing everybody on board.

uture T:avel

Human-made objects have already travelled millions of kilometres into the depths of space. The Pioneer and Voyager space probes have flown beyond Pluto, heading for the edge of the Solar System. However, in the vast expanses of the Universe, this is hardly any distance at all! Scientists and engineers are already working on new technology that will take the human race to the nearest stars, to neighbouring galaxies, and beyond.

Faster-than-light travel

Because the distances between stars are so enormous, we do not possess the technology today to be able to travel to another Solar System. Even the closest star to our Sun is too far away to reach in a single lifetime—Proxima Centauri would take hundreds of thousands of years to reach using the Space Shuttle. To reach other stars, starships have to be able to travel faster than the speed of light—300,000 kilometres (186,000 miles) per second. But according to one of the fundamental laws of physics, Einstein's theory of relativity, faster-than-light travel is impossible. Even if scientists were to find a way to travel at the speed of light, it would still take over 300,000 years to get to the centre of our galaxy.

Interstellar cities

One way of overcoming the need to travel faster than light is to build enormous starships that would act as portable space cities. Some scientists have suggested building colossal constructions, several kilometres long, which would be able to carry thousands of people into space. These ships would not be able to travel as fast as the speed of light, and the journey to find other planets would still take thousands of years. However, the number of people on board would mean that future generations would be able to explore far-off galaxies and planets. None of the original astronauts would live to see the far reaches of space, but their great-great-great grandchildren could!

Einstein's theory of special relativity states that the faster an object moves, the heavier it becomes, so anything travelling at the speed of light would have infinite mass.

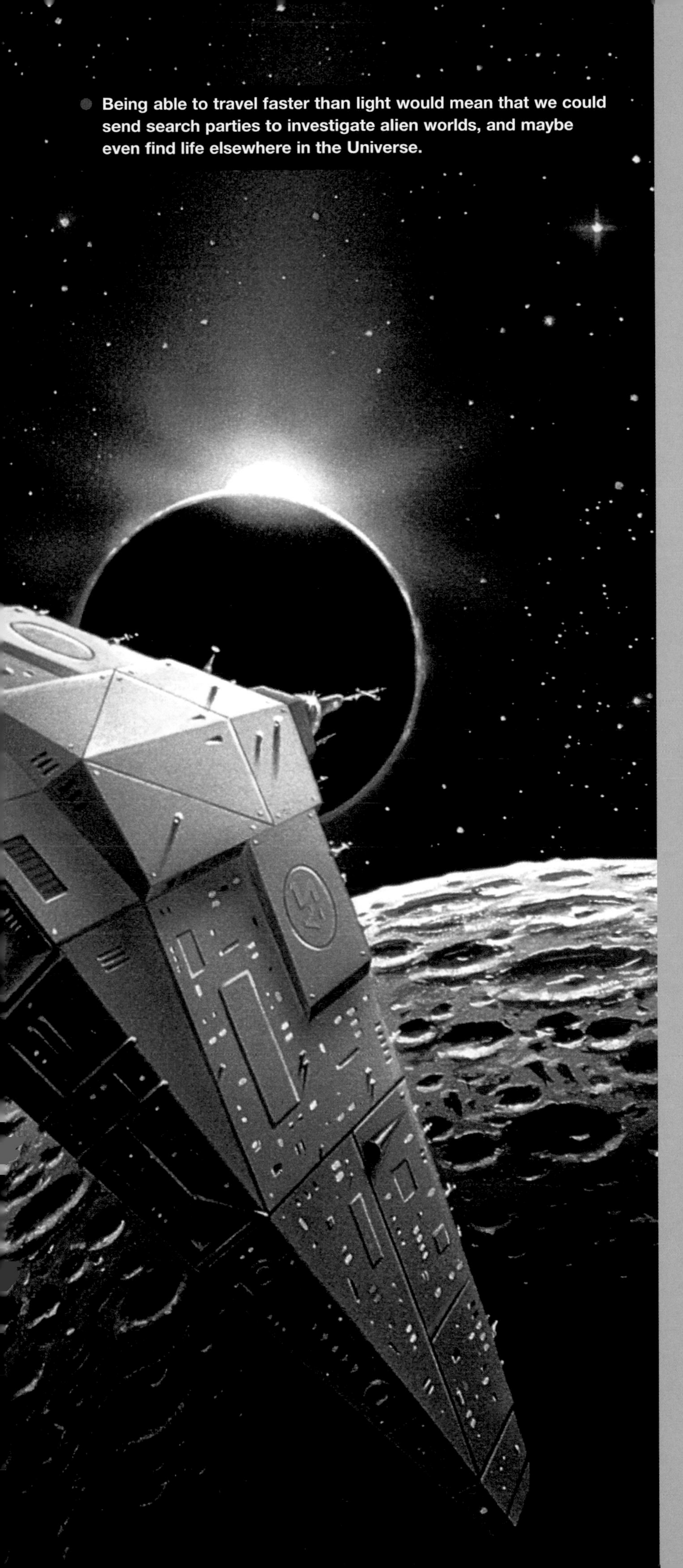

Being able to travel faster than light would mean that we could send search parties to investigate alien worlds, and maybe even find life elsewhere in the Universe.

Future solutions

Scientists have suggested many ways to make starships travel faster than they can today.

Some believe that antimatter engines would be able to propel starships using the vast amounts of energy released when matter and antimatter collide. This would allow space vehicles to travel at much faster speeds but still not at the speed of light.

Another suggestion is to give starships giant sails. The vehicles would be pushed forward by lasers beamed from Earth. The lack of air resistance in space would allow these ships to travel very quickly but not at light speed.

Another method, favoured by science-fiction fans, is to create a warp drive. This would work by contracting space in front of a starship and expanding space behind it. Although the ship itself would not be travelling at the speed of light, the "bubble of space" carrying the ship would be.

Living Off-world

Space stations allow humans to live in space for long periods of time. Cosmonauts lived on Mir for over a year. However, living in Earth's orbit is just the beginning of a new age of space habitation. Scientists all over the world are thinking of new ways to allow people to live in space and on other planets—not just for a year but permanently!

Cities in space

Current space stations are usually quite small and cramped. They cannot hold more than a few people. Even the International Space Station is restricted to a staff of seven. Some scientists have suggested building much larger homes in Earth's orbit. These space cities would not just be laboratories for space science but homes for thousands of people. Although there is no gravity in space, this could be overcome by making the space city spin at a constant speed. This would create a false gravitational force that would push everything to the outer wall.

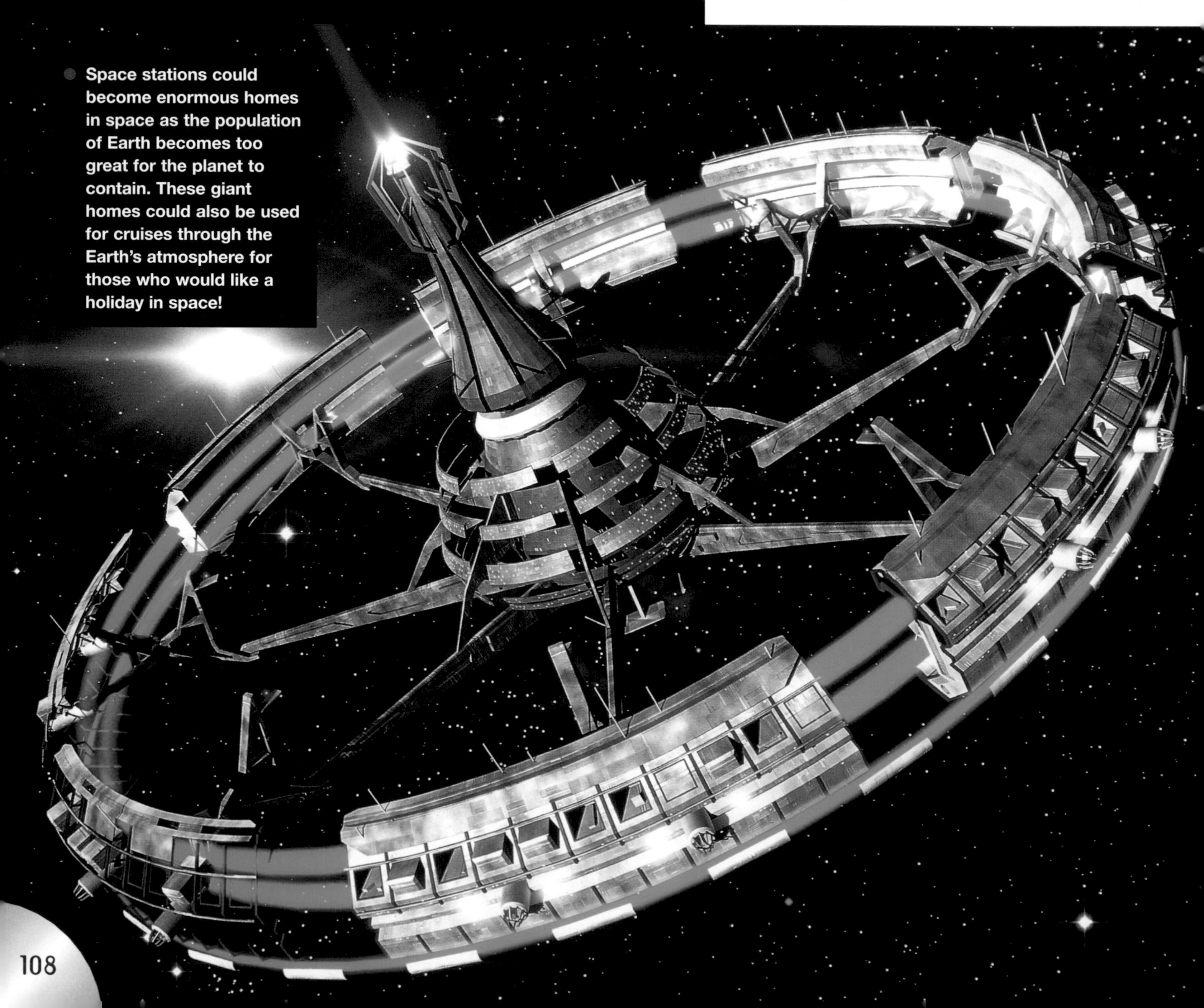

Space stations could become enormous homes in space as the population of Earth becomes too great for the planet to contain. These giant homes could also be used for cruises through the Earth's atmosphere for those who would like a holiday in space!

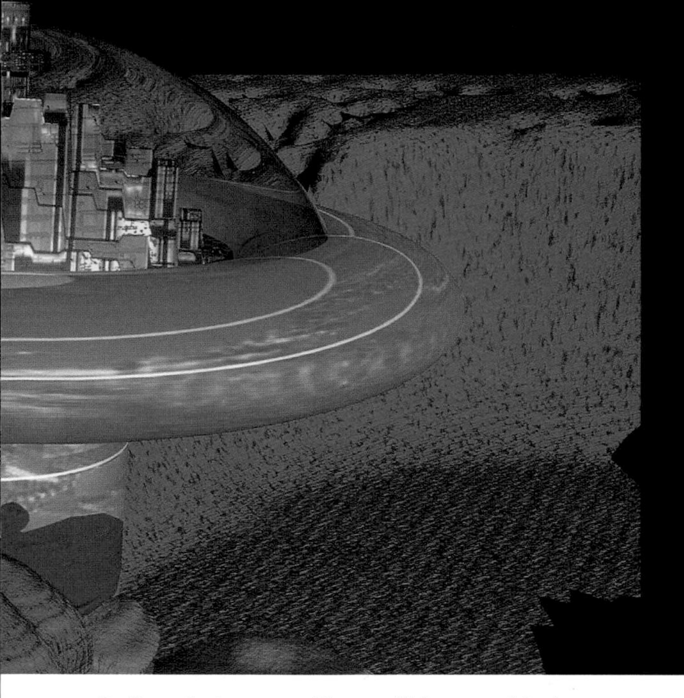

A new home?

Scientists have long been working on a way to transform the atmosphere on Mars to allow people to live there. At the moment, Mars is too cold and the atmosphere too thin to sustain human life. However, a process called terraforming could transform the planet into a new Earth. This would involve unfreezing the planet's polar caps to produce an atmosphere and to allow water to collect on the surface of the planet. After many centuries, the greenhouse effect would enable the temperature and pressure to rise to a level where simple plants could survive. These plants would convert some of the carbon dioxide gas in the atmosphere to oxygen, making the air breathable for human beings.

- **People born on Mars will be used to lower gravity and might not be able to stand Earth's gravity. The first Martians to land on Earth may be human!**
- **Because of the number of people on board a future space city, crops would have to be grown inside the station, and many things, including human waste, would have to be recycled!**
- **If human beings do ever arrive on an alien planet, who knows what sort of building materials they may find? There may be certain kinds of metals that are easier to use than those on Earth, enabling vast cities, like the one below, to be built.**

Moon laboratories

Astronomy from Earth can be difficult because of our planet's turbulent atmosphere. Telescopes in orbit, like Hubble, allow astronomers a much clearer view of space but cannot be as easily maintained as telescopes on Earth. Within fifty years, astronomers plan to return to the Moon to build permanent bases there. The Moon has no atmosphere, which makes it an ideal place for an observatory. It also has very low gravity, which means that enormous satellites and telescopes can be built much more easily than they could be on Earth. Astonomers will live on the Moon to carry out their experiments.

A Career In Space

Would you like to be at the helm of a Space Shuttle as it blasts into orbit? Or be the astronomer who discovers the ninth planet in our Solar System? Or be the first person to set foot on Mars? Technology is advancing at an incredible rate. Before long it could be you who is at the forefront of space exploration. Work hard and stay healthy, and space could be your next destination!

- Imagine being the first astronomer to pick up the sounds of an alien message! You may even be asked to greet the aliens if they arrive on Earth.
- To be a pilot astronaut you must have 1,000 hours flight time in a jet aircraft. But because we now know much more about space travel, most astronauts are no longer pilots but scientists.

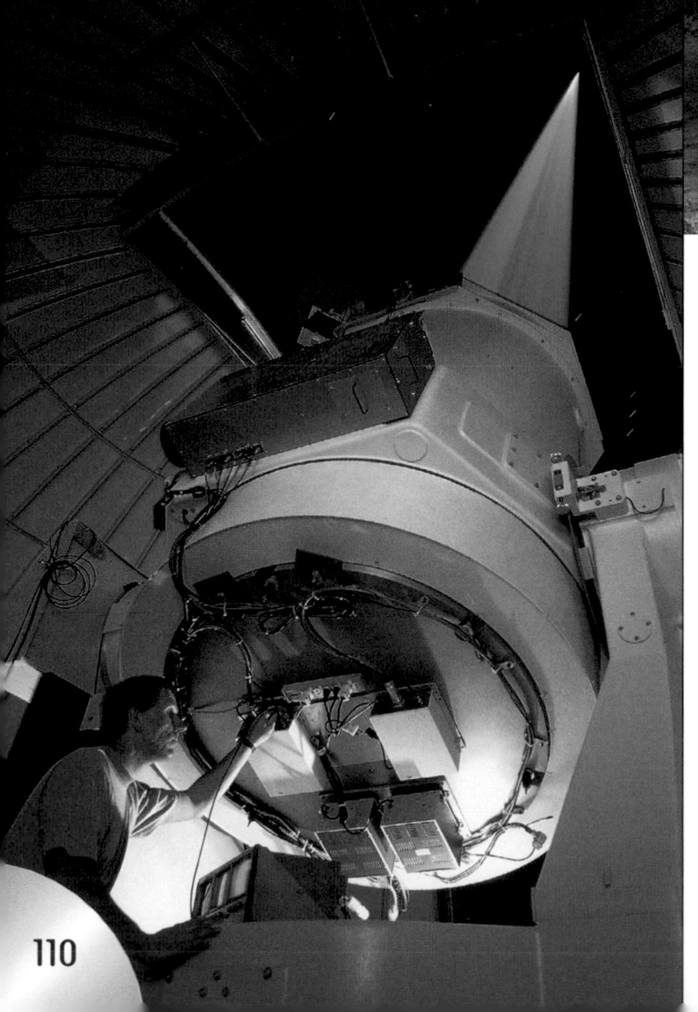

Astronomy as a career

Do you fancy analysing the Hubble Space Telescope's findings rather than exploring space from your back garden? If so, then the job of a professional astronomer could be for you. All the major space agencies across the globe employ professional astronomers to analyse the data that is collected by telescopes and space probes. This is not always done by looking at space, however. Many professional astronomers may not even look at the sky! Instead, they work with computers and complicated mathematics to help explain the mysteries of the Universe. Because of this, it is important to have a good knowledge of maths and physics to become a professional astronomer.

Space Inc.

The excitement of space exploration is what prompts many young scientists to work towards a job in astronomy, but some entrepreneurs have already found inventive ways to make money from the extraordinary possibilities of the Universe.

- **A private company has sent a number of tourists into space, and other space tourism companies are taking reservations for future trips.**
- **Other organizations offer the chance to have your mortal remains sent into space after your death—at a price.**
- **It is possible to pay to have a star named after yourself or a loved one. However, astronomers themselves are likely to continue to use the official names and numbers they have allocated to stars.**

Become an astronaut

For many, the most exciting job in space exploration is that of an astronaut—to pilot the Space Shuttle as it deploys a satellite, or to be a mission specialist on the first manned mission to another planet. It is never too early to start training to be an astronaut. Work hard at school, especially in maths and physics. Look after your body by keeping fit and healthy. To become a pilot astronaut with most space agencies, you will probably need some experience flying jet aircraft. But there are many other jobs in space exploration. Astronauts would not even get into space without the help of thousands of other men and women.

Astronomy from Home

Billions are spent by governments around the world on looking at the stars and the planets. Professional astronomers may never even look at the sky itself but instead use computers and complicated mathematics to help explain the mysteries of space. However, none of this is necessary to view the fascinating beauty of the sky. Thousands of people worldwide marvel at the wonders of space using small telescopes or even the naked eye.

Getting familiar

It may seem daunting when you gaze up at the thousands of stars visible in the night sky. However, it is worth spending the time familiarizing yourself with the patterns of the stars and the positions of the planets. Use the star maps on pages 120 and 121 to help you find your way around the night sky. Once you can pick out the constellations, you will be able to use them to locate the planets. Venus, Mars, Jupiter and Saturn can usually be found close to the Sun's path through the sky. If you see a bright "star" that you cannot identify on the star map, it is likely to be a planet.

Where to start

If the sky is clear tonight, you can begin skywatching immediately. If you have a pair of binoculars or a small telescope, you will be able spot faint stars, see some of the planets and even pick out craters on the Moon. Naked-eye astronomy can be just as exciting. See if you can pick out the constellations and galaxies, bright nebulae and, from time to time, lunar eclipses. The night sky is best observed from a location away from built-up areas, because light pollution can often drown out the light of fainter stars. The light from the full Moon can also make it difficult to spot some objects in the night sky. Remember to dress warmly when you are outside at night, and you should always be accompanied by an adult when you are star-spotting.

- **See if you can spot any of the constellations. Stars and planets remain in almost the same position each night.**
- **Brightly coloured nebulae can often be seen with binoculars.**
- **Meteors, or shooting stars, are brilliant streaks of light that usually last for fractions of a second.**
- **If you see a bright star near the horizon that you cannot identify on the star map, it is probably Venus.**

Useful items

- A torch covered with red film so that it does not affect your night vision.

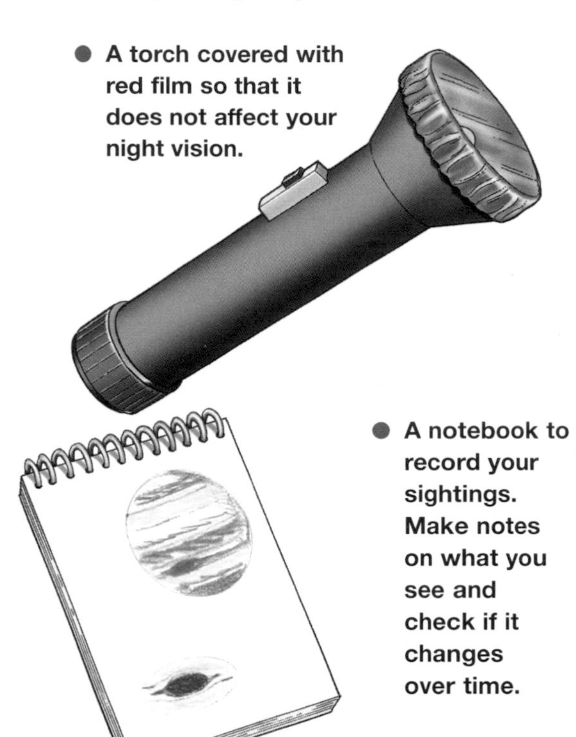

- A notebook to record your sightings. Make notes on what you see and check if it changes over time.

- If you are very lucky, you may be able to see a comet. They can appear for days or weeks at a time.
- The Moon can be seen in different positions each night. When it is full, use a telescope to spot large craters.
- Some satellites can be seen as bright flashes of light.

Astronomy by day

Astronomy does not always have to be done at night. Some celestial objects can also be seen during the day, including the Moon, Venus and, of course, the Sun. Remember it is very dangerous to look directly at the Sun. However, you can observe the Sun safely by projecting its image onto card through a telescope or pair of binoculars. Using the method below you can watch solar eclipses as they happen, but make sure you always have an adult with you.

- Measure one of the front lenses of your binoculars, and ask an adult to cut a similar sized hole in a large sheet of card.

- Tape the card to the binoculars so that light can shine through one end. The sheet of card stops too much sunlight getting through, which would spoil the image.
- Make a screen from a large piece of white card and hold the binoculars 50 centimetres (20 inches) away so that the disk of the Sun can be seen projected on the card. Focus the binoculars until the image is sharp.

How many Earths could fit into the giant planet Jupiter? How hot is the surface of the Sun? Which planet has the most moons? Where in the night sky is the Hercules constellation? How many rings does Neptune have? What is the closest star to our Sun?

Looking at space can leave you with many mind-boggling questions that can be answered by looking at the information provided in this section. There are star charts to help you navigate your way around the night sky and a glossary to remind you of some of the terms that are used in this book.

Reference Section

Planets

	Mercury	Venus	Earth	Mars	Jupiter	Saturn	Uranus	Neptune
Equatorial diameter (km):	4,877	12,100	12,756	6,785	142,984	120,536	51,166	49,557
Distance from Sun (million km):	58	108	150	228	779	1,427	2,869	4,496
Length of orbit (Earth days):	88	225	365	687	4,333	10,760	30,065	60,190
Speed of orbit (km/h):	172,332	126,072	107,244	86,868	47,016	34,812	24,516	19,548
Length of day (Earth hours):	1,408	5,832	24	25	10	10	18	19
Mass (x Earth):	0.055	0.815	1	0.107	317.9	95.2	14.5	17.1
Density (Water=1):	5.42	5.25	5.52	3.94	1.33	0.69	1.27	1.71
Surface gravity (x Earth):	0.38	0.9	1	0.38	2.36	0.92	0.89	1.13
Surface temperature (°C):	170	480	15	–63	–121	–180	–195	–200
Number of moons:	0	0	1	2	16	18	17	8
Number of rings:	0	0	0	0	3	7	11	6
Time taken to form (Earth years):	80,000	40,000	110,000	200,000	1 million	9 million	300 million	1 billion

Moons

Planet	Name	Diameter (km)	Distance from planet (km)	Time taken to orbit planet (Earth days)
Earth	Moon	3,476	384,400	27.3
Mars	Phobos	22	9,400	0.3
	Deimos	13	23,500	1.4
Jupiter	Metis	40	128,000	0.3
	Adrastea	20	129,000	0.3
	Amalthea	200	181,300	0.5
	Thebe	100	221,900	0.7
	Io	3,640	421,800	1.8
	Europa	3,138	670,900	3.6
	Ganymede	5,262	1,070,000	7.2
	Callisto	4,800	1,880,000	16.7
	Leda	15	11,094,000	238.7
	Himalia	170	11,480,000	250.6
	Lysithea	35	11,720,000	259.2
	Elara	70	11,737,000	259
	Ananke	25	21,200,000	631
	Carme	40	22,600,000	692
	Pasiphae	60	23,500,000	735
	Sinope	40	23,700,000	758
Saturn	Pan	20	133,600	0.57
	Atlas	31	137,700	0.6
	Prometheus	102	139,400	0.6
	Pandora	85	141,700	0.6
	Epimetheus	117	151,400	0.7
	Janus	188	151,500	0.7
	Mimas	397	186,000	0.9
	Enceladus	498	238,000	1.4
	Tethys	1,050	295,000	1.9
	Telesto	22	295,000	1.9
	Calypso	24	295,000	1.9
	Dione	1,118	377,000	2.7
	Helene	32	377,000	2.7
	Rhea	1,528	527,000	4.6
	Titan	5,150	1,222,000	15.9
	Hyperion	286	1,481,100	21.3
	Iapetus	1,436	3,561,300	79.3
	Phoebe	220	12,954,000	550.4
Uranus	Cordelia	26	49,700	0.3
	Ophelia	32	53,800	0.4
	Bianca	44	59,200	0.4
	Cressida	66	61,800	0.5
	Desdemona	58	62,700	0.5
	Juliet	84	64,400	0.5
	Portia	110	66,100	0.5
	Rosalind	58	69,900	0.6
	Belinda	68	75,300	0.6
	Puck	154	86,000	0.8
	Miranda	472	129,800	1.4
	Ariel	1,158	191,200	2.5
	Umbriel	1,169	266,000	4.1
	Titania	1,578	435,900	8.7
	Oberon	1,123	582,600	13.5
Neptune	Naiad	54	48,000	0.3
	Thalassa	80	50,000	0.3
	Despina	180	52,500	0.3
	Galatea	150	62,000	0.4
	Larissa	192	73,600	0.6
	Proteus	416	117,600	1.1
	Triton	2,705	354,800	5.9
	Nereid	300	5,514,000	360.2

Io

Moon

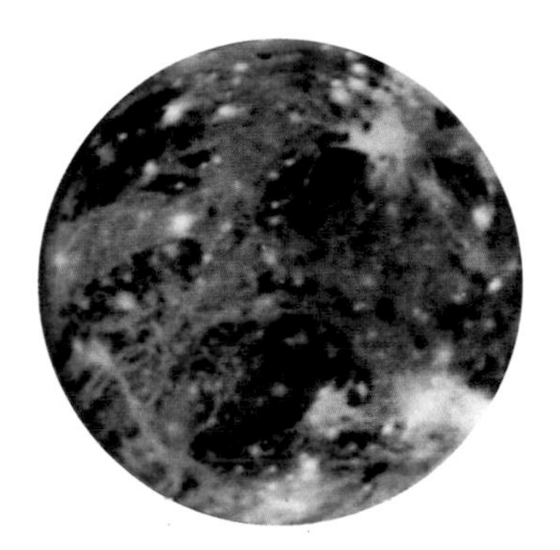
Ganymede

Titan

Callisto

The Sun

Equatorial diameter (million km)	1.4
Diameter (x Earth)	109
Mass (x Earth)	330,000
Density (Water = 1)	1.4
Average distance from Earth (million km)	150
Rotation period (Earth hours)	610–864
Composition	73% hydrogen 25% helium 2% other
Surface temperature (°C)	5,500°C
Core Temperature (°C)	15,000,000°C
Age (Earth years)	4.6 billion
Luminosity (megawatts):	390 billion billion

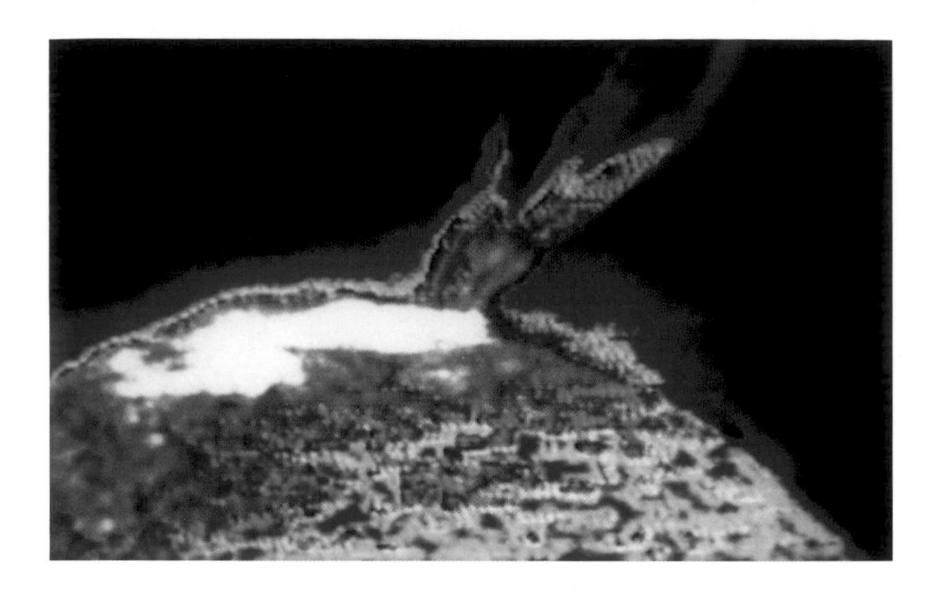

Nearest Stars

Name	Constellation	Distance (light years)
Sun	—	0.000015
Proxima Centauri	Centaurus	4.2
Alpha Centauri A, B and C	Centaurus	4.3
Barnard's Star	Ophiuchus	6.0
Wolf 359	Leo	7.7
Lalande 21185	Ursa Major	8.2
Sirius A	Canis Major	8.6
Sirius B	Canis Major	8.6
UV Ceti A	Cetus	8.9
Ross 154	Sagittarius	9.6

Brightest Stars

Name	Constellation	Magnitude	Distance (ly)
Sun	—	–26.7	0.000015
Sirius A	Canis Major	–1.46	8.6
Canopus	Carina	–0.72	313
Alpha Centauri	Centauris	–0.27	4.3
Arcturus	Boötes	–0.04	36
Vega	Lyra	+0.03	25
Capella	Auriga	+0.08	42
Rigel	Orion	+0.12	773
Procyon	Canis Major	+0.38	11
Achernar	Endanus	+0.46	144

Constellations

Name	Common name
Andromeda	Andromeda
Antlia	Air Pump
Apus	Bird of Paradise
Aquarius	Water Carrier
Aquila	Eagle
Ara	Altar
Aries	Ram
Auriga	Charioteer
Boötes	Herdsman
Caelum	Chisel
Camelopardalis	Giraffe
Cancer	Crab
Canes Venatici	Hunting Dogs
Canis Major	Great Dog
Canis Minor	Little Dog
Capricornus	Sea Goat
Carina	Keel
Cassiopeia	Cassiopeia
Centaurus	Centaur
Cepheus	Cepheus
Cetus	Whale
Chamaeleon	Chamaeleon
Circinus	Compasses
Columba	Dove
Coma Berenices	Berenice's Hair
Corona Australis	Southern Crown
Corona Borealis	Northern Crown
Corvus	Crow
Crater	Cup
Crux	Southern Cross
Cygnus	Swan
Delphinus	Dolphin
Dorado	Swordfish
Draco	Dragon
Equuleus	Foal
Eridanus	River
Fornax	Furnace
Gemini	Twins
Grus	Crane
Hercules	Hercules
Horologium	Clock
Hydra	Water Snake
Hydrus	Little Water Snake
Indus	Indian
Lacerta	Lizard
Leo	Lion
Leo Minor	Little Lion
Lepus	Hare
Libra	Scales
Lupus	Wolf
Lynx	Lynx
Lyra	Lyre
Mensa	Table Mountain
Microscopium	Microscope
Monocerus	Unicorn
Musca	Fly
Norma	Level
Octans	Octant
Phiuchus	Serpent Bearer
Orion	Hunter
Pavo	Peacock
Pegasus	Winged Horse
Perseus	Perseus
Phoenix	Phoenix
Pictor	Painter's Easel
Pisces	Fish
Piscis Austrinus	Southern Fish
Puppis	Stern
Pyxis	Mariner's Compass
Reticulum	Net
Sagitta	Arrow
Sagittarius	Archer
Scorpius	Scorpion
Sculptor	Sculptor
Scutum	Shield
Serpens	Serpent
Sextans	Sextant
Taurus	Bull
Telescopium	Telescope
Triangulum	Triangle
Triangulum Australe	Southern Triangle
Tucana	Toucan
Ursa Major	Great Bear
Ursa Minor	Little Bear
Vela	Sails
Virgo	Virgin
Volans	Flying Fish
Vulpecula	Fox

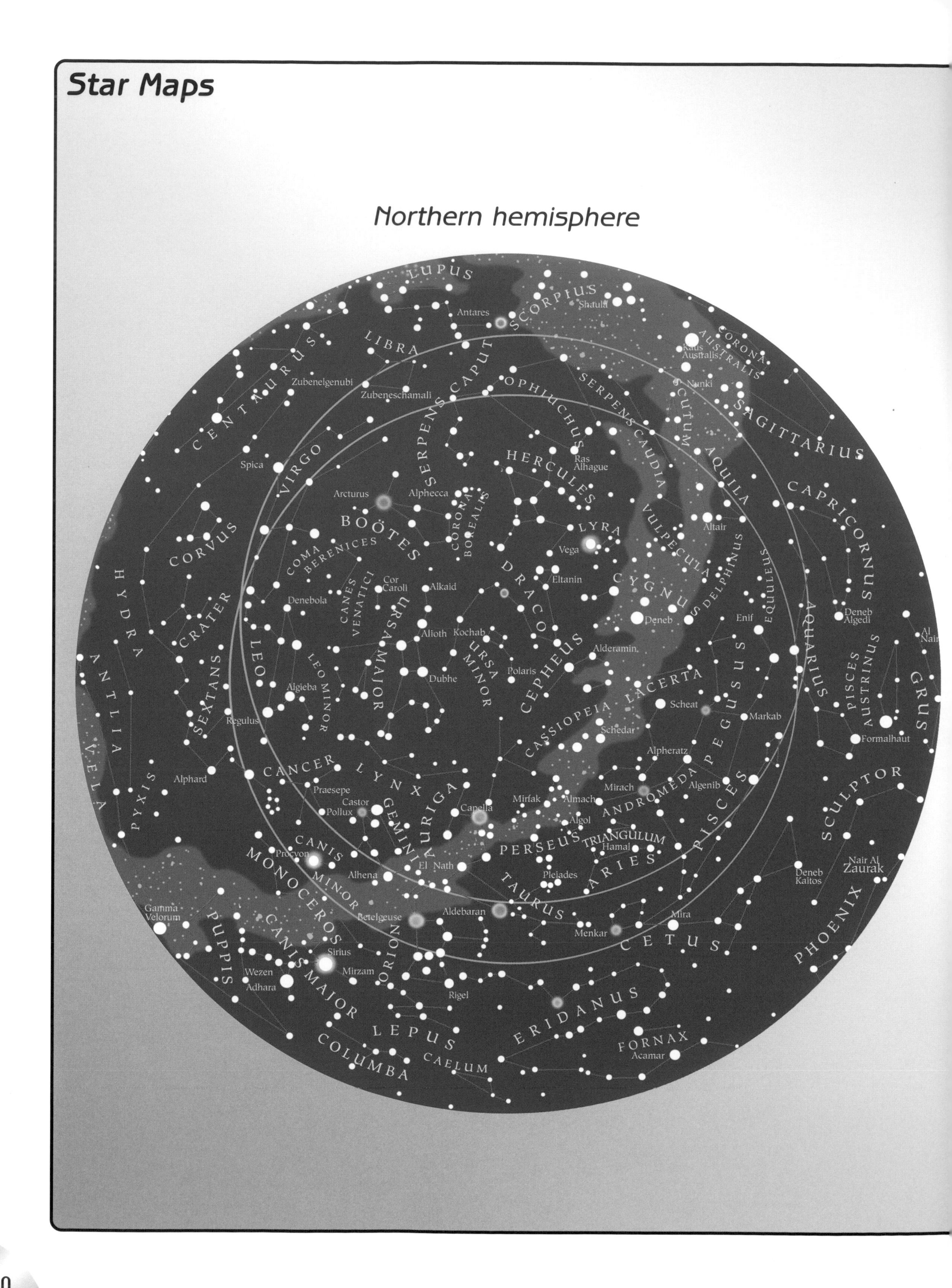
Star Maps
Northern hemisphere
LUPUS
SCORPIUS
Shaula
Antares
LIBRA
CORONA AUSTRALIS
Kaus Australis
Nunki
SAGITTARIUS
CENTAURUS
Zubenelgenubi
Zubeneschamali
SERPENS CAPUT
OPHIUCHUS
SERPENS CAUDA
SCUTUM
AQUILA
Spica
VIRGO
Ras Alhague
HERCULES
CAPRICORNUS
Arcturus
Alphecca
CORONA BOREALIS
LYRA
Altair
CORVUS
BOÖTES
COMA BERENICES
Vega
VULPECULA
DELPHINUS
EQUULEUS
HYDRA
CANES VENATICI
Cor Caroli
Alkaid
DRACO
Eltanin
CYGNUS
Deneb
Enif
Deneb Algedi
CRATER
Denebola
URSA MAJOR
Alioth
Kochab
URSA MINOR
AQUARIUS
Al Nair
ANTLIA
LEO
LEO MINOR
Dubhe
Polaris
CEPHEUS
Alderamin
PEGASUS
PISCES AUSTRINUS
GRUS
SEXTANS
Algieba
Regulus
CASSIOPEIA
LACERTA
Scheat
Markab
Schedar
Formalhaut
VELA
Alpheratz
PYXIS
Alphard
CANCER
LYNX
Praesepe
Castor
Pollux
GEMINI
AURIGA
Capella
Mirfak
Almach
Mirach
ANDROMEDA
Algenib
PISCES
SCULPTOR
Algol
TRIANGULUM
Hamal
CANIS MINOR
Procyon
MONOCEROS
El Nath
PERSEUS
ARIES
Alhena
Pleiades
Deneb Kaitos
Nair Al Zaurak
Gamma Velorum
TAURUS
Betelgeuse
Aldebaran
Mira
Menkar
PHOENIX
PUPPIS
CANIS MAJOR
Sirius
ORION
CETUS
Wezen
Adhara
Mirzam
Rigel
ERIDANUS
LEPUS
FORNAX
Acamar
COLUMBA
CAELUM

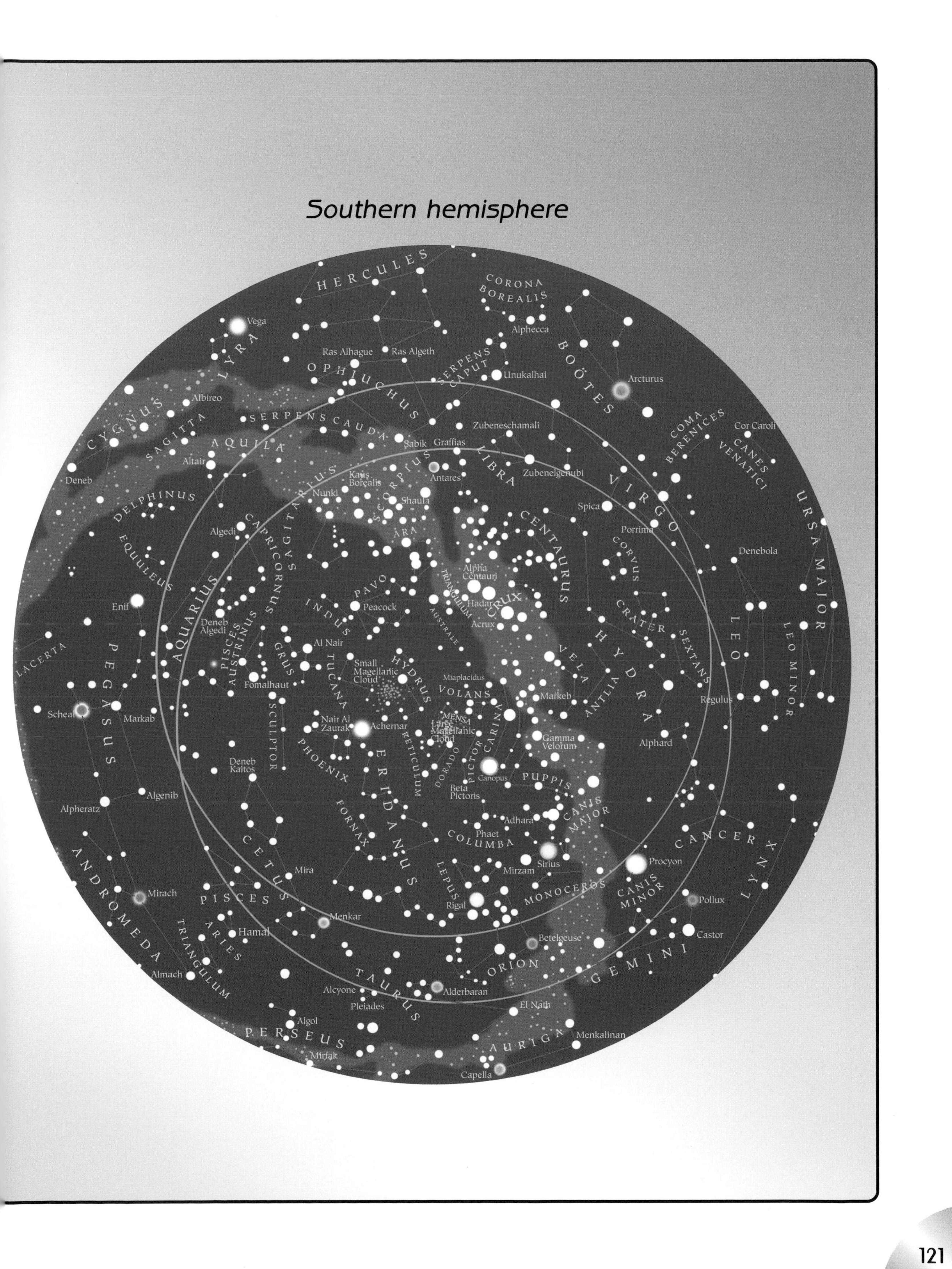

Southern hemisphere
HERCULES
CORONA BOREALIS
Alphecca
Vega
LYRA
Ras Alhague
Ras Algeth
SERPENS CAPUT
Unukalhai
BOÖTES
Arcturus
OPHIUCHUS
Albireo
CYGNUS
SAGITTA
SERPENS CAUDA
Zubeneschamali
COMA BERENICES
Cor Caroli
CANES VENATICI
AQUILA
Sabik
Graffias
LIBRA
Altair
Deneb
Kaus Borealis
Antares
Zubenelgenubi
VIRGO
Nunki
SCORPIUS
Shaula
DELPHINUS
SAGITTARIUS
Spica
CENTAURUS
URSA MAJOR
Algedi
CAPRICORNUS
ARA
Porrima
CORVUS
EQUULEUS
Denebola
Alpha Centauri
TRIANGULUM AUSTRALE
Hadar
CRUX
PAVO
INDUS
Peacock
Enif
AQUARIUS
Deneb Algedi
Acrux
CRATER
LEO
LEO MINOR
SEXTANS
PISCES AUSTRINUS
Al Nair
GRUS
HYDRA
LACERTA
PEGASUS
TUCANA
Small Magellanic Cloud
HYDRUS
VELA
Miaplacidus
Fomalhaut
VOLANS
ANTLIA
Markeb
Regulus
Scheat
Markab
MENSA
Large Magellanic Cloud
CARINA
Nair Al Zaurak
Achernar
RETICULUM
SCULPTOR
Gamma Velorum
Alphard
PHOENIX
ERIDANUS
DORADO
PICTOR
Deneb Kaitos
Canopus
PUPPIS
Beta Pictoris
Algenib
Alpheratz
FORNAX
CANIS MAJOR
Adhara
Phaet
COLUMBA
CANCER
LYNX
CETUS
Sirius
Procyon
ANDROMEDA
Mira
Mirzam
LEPUS
MONOCEROS
CANIS MINOR
Mirach
PISCES
Rigal
Pollux
Menkar
TRIANGULUM
ARIES
Hamal
Betelgeuse
Castor
ORION
GEMINI
Almach
TAURUS
Alcyone
Pleiades
Alderbaran
El Nath
Algol
PERSEUS
AURIGA
Menkalinan
Mirfak
Capella

Glossary

Absolute zero The coldest temperature possible, or –273°C (–459°F).

Accretion disk A flat disk of gas surrounding a black hole or a newborn star.

Active galaxy A galaxy that generates a huge amount of energy because of a black hole at its centre.

Annular eclipse A solar eclipse where only the outer rim of the Sun is left uncovered by the Moon, leaving a "ring of fire".

Antimatter Matter made from subatomic particles that acts in an opposite way to normal matter, and which can annihilate ordinary particles.

Asteroid A small, rocky body orbiting the Sun. Asteroids can range in size from smaller than a few metres in diameter to larger than a thousand kilometres across.

Astrology The belief that the movement of the Sun, the Moon, the stars and the planets affects the lives of human beings.

Astronomical unit (AU) The average distance between the Sun and the Earth, approximately 93 million million miles (150 million million kilometres).

Atmosphere A layer of gases around a celestial object such as a planet or moon, held there by the object's gravity.

Atom The smallest part of an element, made up of subatomic particles called protons, neutrons and electrons.

Aurora Coloured lights in the skies above the polar regions, caused by particles from the Sun hitting gases in the Earth's atmosphere and causing them to glow.

Axis The imaginary line through the centre of a celestial object such as a planet or a moon, around which it rotates.

Background radiation A faint radio signal emitted from the whole sky, thought to be created by the Big Bang.

Big Bang A theory of the beginning of the Universe, which states that it started from a violent explosion around 13 billion years ago.

Billion One thousand million.

Black hole The core of a collapsed star with a gravitational pull so strong that nothing can escape from it, not even light.

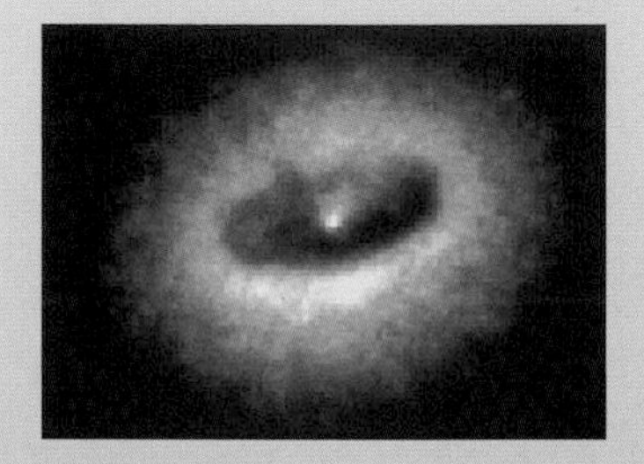

Brown dwarf A star that is smaller than a planet, producing heat but not light.

Celestial object Any object in the sky, including stars, planets and moons.

Celestial sphere An imaginary sphere around the Earth, on which the Sun, the Moon, the planets and the stars all appear to be fixed.

Chromosphere The lower region of the Sun's atmosphere, between the photosphere and the corona.

Combustion The burning of fuel in an engine to provide power. Used in rocket engines.

Comet An object composed of ice and dust that orbits the Sun on an elliptical orbit. When it travels too close to the Sun, the solar wind melts the ice in a comet, forming a tail.

Constellation One of the 88 official patterns of stars in the night sky, many named after mythical characters and animals.

Corona The very hot upper atmosphere of the Sun, visible only during a solar eclipse.

Crater A large dent in the surface of a planet or moon, caused by the powerful impact of a meteorite.

Crust The rocky outer layer of a planet or moon.

Dark matter Invisible matter, including brown dwarfs, which is thought to make up over 95% of the mass of the Universe.

Dual planet system A term used to describe a pair of planets, or a planet and a moon, that orbit around each other.

Ecliptic An imaginary line across the sky, along which the Sun appears to move. It is actually a projection of the Earth's orbit around the Sun onto the sky.

Ecosphere The narrow band around the Sun, or another star, in which it is neither too hot nor too cold for water to exist in liquid form and thus for life to begin.

Electromagnetic radiation Waves of energy that travel at the speed of light. It is found in different wavelengths, from gamma rays, which have the shortest wavelength, to radio waves, which have the longest wavelength.

Element A basic substance with unique properties, which cannot be broken down by a chemical reaction.

Elliptical A term used to describe an elongated circle. Elliptical galaxies are oval in shape and made up of old stars and a little gas. They have no spiral arms.

Escape velocity The minimum speed at which an object must travel to escape the gravity of a planet or a moon. The escape velocity needed to break free of Earth's surface is about 40,000 km/h (25,000 mph).

Extrasolar Anything that exists beyond the limits of our Solar System.

Extraterrestrial Anything that isn't from the Earth, usually applied to alien life.

Frequency The number of waves of electromagnetic radiation that pass a certain point every second.

Galaxy An enormous body of stars, gas and dust that is held together by gravity. Galaxies can be spiral, elliptical or irregular in shape, and are separated from each other by voids of empty space.

Gas giant An enormous planet made up of a deep atmosphere of gas with a solid core.

Gravity The force of attraction between any two objects with mass. Every object has a gravitational pull, but with most objects it is so small it is hardly noticeable. Planets, stars and black holes all have much more mass and therefore have a much stronger gravitational pull.

Greenhouse effect The warming of a planet's surface when the Sun's radiation is trapped by gases in the atmosphere, in the same way as the glass of a greenhouse prevents heat from escaping.

Hertzsprung–Russell diagram A diagram showing the relationship between a star's colour and its brightness. It shows how stars can be organized into a few main types, and helps us understand their life cycle.

Hydrogen The simplest and most common element in the Universe and the main element in stars. Nuclear fusion in stars turns hydrogen into helium.

Infrared The part of the electromagnetic spectrum between visible light waves and radio waves. It is also known as heat radiation.

Ion An electrically charged atom.

Irregular galaxy A galaxy with no recognizable shape, made up of a mixture of old and new stars.

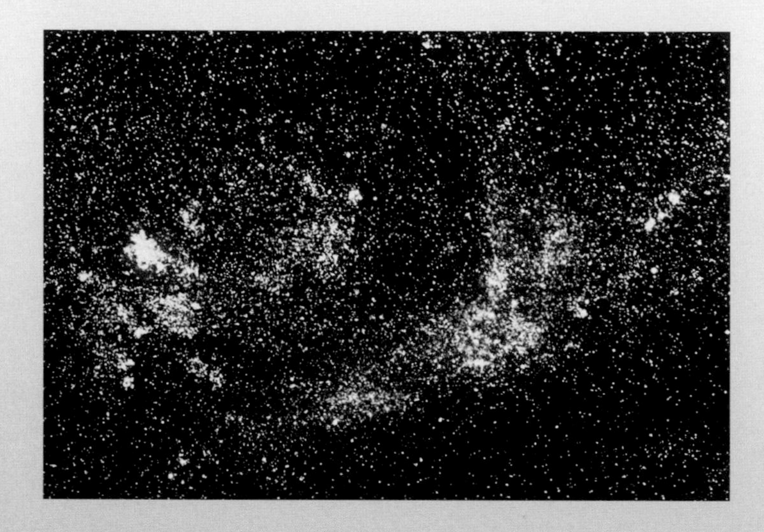

Lava Molten rock released from the interior of a planet, usually through a volcano.

Light Energy in the form of electromagnetic radiation, with wavelengths that are visible to the human eye. Light is the fastest thing in the Universe.

Light year A unit of measurement based on the distance that light travels in a year, roughly 9.5 million million kilometres (6 million million miles).

Local Arm The spiral arm of the Milky Way in which our Solar System lies, also known as the Orion arm.

Local Group The cluster of galaxies to which the Milky Way belongs.

Luminosity The total brightness of a star.

Magnetic field The magnetic force generated by a celestial body.

Magnitude A unit used to measure the brightness of celestial objects. Bright objects have low magnitudes, and less bright objects have higher magnitudes.

Main sequence The band on the Hertzsprung–Russell diagram where stars lie for most of their lives as they convert hydrogen into helium.

Mass How much matter there is in an object, and how this matter is influenced by gravity.

Matter Anything that has mass and occupies space.

Meteoroid Fragments of dust and rock in space, usually from asteroids and comets. If a meteoroid enters the Earth's atmosphere, it burns up and becomes a meteor. If it lands on the Earth's surface, it is called a meteorite.

Microgravity Extremely low gravity. This is what astronauts experience when they float in orbit, as there are no objects with large enough gravitational forces to pull them to any surface.

Microwaves Part of the electromagnetic spectrum. Microwaves are radio waves with the shortest radio wavelengths.

Milky Way The name given to the galaxy in which our Solar System lies. It can be seen as a band of stars across the night sky.

Million A thousand thousand.

Molecule A group of atoms linked together by chemical bonds.

Moon Any natural satellite orbiting a planet.

Nebula A cloud of gas and dust in space that is visible either when it reflects light from nearby stars, or when it blocks out light from more distant stars.

Nuclear fusion The source of energy for all stars. Energy is released when atoms are joined together to form new elements.

Orbit The path of one celestial object around another celestial object with greater mass. Satellites and planets are examples of things held in orbit around a more massive object because of its gravitational pull.

Organism A living creature made from organic cells.

Oxygen An element that makes up 20% of the gas in the Earth's atmosphere. Life on Earth could not exist without it.

Planet A spherical celestial object that orbits a star. Planets can be made of rock or gas but do not produce their own light.

Planetisimals Minute pieces of rock that joined together during the beginnings of the Solar System to form the planets.

Pressure The amount of force on a specific area. Air pressure is the force with which the air presses down onto the surface of a planet or moon.

Propulsion When an object is pushed or driven forwards. Conventional rocket technology uses the exhaust gases produced by mixing liquid hydrogen and liquid oxygen to propel rockets into orbit.

Pulsar A neutron star that spins rapidly, beaming waves of radiation across space like a lighthouse beaming light.

Quasar A very distant active galaxy that releases huge amounts of energy from a black hole in its centre.

Satellite Any object held in the gravitational pull of a larger object. A moon orbiting a planet is a satellite, as is an artificial satellite that orbits the Earth. Small galaxies can also be satellites around larger ones.

Singularity When all matter exists in a minuscule point of infinite density. Scientists believe that all matter existed in a singularity before it was released in the Big Bang. Also, anything pulled into a black hole is compressed into a singularity.

Spiral galaxy A galaxy with a central hub from which emerge spiral arms. Spiral galaxies are usually made up of old and young stars, and large amounts of dust.

Star A ball of gas that generates energy by nuclear fusion.

Terraforming The process of changing a planet's atmosphere to make it habitable for humans. This would involve creating an atmosphere to trap the Sun's radiation.

Trillion A million million.

Universe Everything that exists.

Zodiac The twelve constellations through which the Sun appeared to early astronomers to move. The Zodiac is still used by astrologers, who believe the position of the stars can affect human lives.

Index